A First Course in Ordinary Differential Equations

Martin Hermann • Masoud Saravi

A First Course in Ordinary Differential Equations

Analytical and Numerical Methods

 Springer

Martin Hermann
Institute of Applied Mathematics
Friedrich Schiller University
Jena, Germany

Masoud Saravi
Department of Mathematics
Islamic Azad University
Nour Branch
Nour, Iran

ISBN 978-81-322-3527-9   ISBN 978-81-322-1835-7 (eBook)
DOI 10.1007/978-81-322-1835-7
Springer New Delhi Heidelberg New York Dordrecht London

Printed on acid-free paper

Springer is part of Springer Science+Business Media (www.springer.com)

To our wives,
Gudrun and Mahnaz,
and our children,
Alexander, Sina, and Sayeh

Preface

This book presents a modern introduction to mathematical techniques for solving initial and boundary value problems in linear ordinary differential equations (ODEs). The focus on analytical and numerical methods makes this book particularly attractive.

ODEs play an important role in modeling complex scientific, technological, and economic processes and phenomena. Therefore, to address students and scientists of various disciplines, we have circumvented the traditional definition-theorem-proof format. Instead, we describe the mathematical background by means of a variety of problems, examples, and exercises ranging from the elementary to the challenging problems.

The book is intended as a primary text for courses in theory of ODEs and numerical treatment of ODEs for advanced undergraduate and early graduate students. It is assumed that the reader has a basic knowledge of elementary calculus, in particular methods of integration, and numerical analysis. Physicists, chemists, biologists, computer scientists, and engineers whose work involves the solution of ODEs will also find the book useful both as a reference and as a tool for self-studying. The book has been prepared within the scope of a German-Iranian research project on mathematical methods for ODEs, which was started in early 2012.

We now outline the contents of the book. In Chap. 1, an introduction to ODEs and some basic concepts are presented. Chapter 2 deals with scalar first-order ODEs. The existence and uniqueness of solutions is studied on the basis of the well-known theorems of Peano and Picard-Lindelöf. The analytical standard techniques for the determination of the exact solution are given. Moreover, the solution of high-degree first-order ODEs (Clairaut and Lagrange equation), which are often encountered in the applications, is discussed. This chapter ends with the discussion about the family of curves and orthogonal trajectories. Chapter 3 is devoted to analytical methods for the solution of second-order ODEs. Subsequent to the introduction to higher-order ODEs, solution methods for homogeneous and inhomogeneous equations are presented. Chapter 4 focuses on the Laplace transform for scalar ODEs that is also used in Chap. 5 for the solution of systems of first-order ODEs.

The Laplace transform is a widely applied integral transform in mathematics used to switch a function from the time domain to the s-domain. The Laplace transform can be used in some cases to solve linear ODEs with given initial conditions. At the end of Chap. 5, it is shown how the solution for the homogeneous system can be determined on the basis of the eigenvalues of the characteristic equation. Then, the corresponding solution for the inhomogeneous system is developed by the method of variation of parameters, and the solution for the ODE is presented by the superposition principle. In Chap. 6, it is shown how power series techniques can be used to represent the solution for scalar first- and second-order ODEs. Special attention is paid to Legendre's equation, Bessel's equation, and the hypergeometric equation since these equations often occur in the applications. The numerical part of this book begins with Chap. 7. Here, numerical methods for initial value problems of systems of first-order ODEs are studied. Starting with the concept of discretizing ODEs, the class of Runge–Kutta methods is introduced. The Butcher schemes of a variety of Runge–Kutta methods are given. Further topics discussed are consistency, convergence, estimation of the local discretization error, step-size control, A-stability, and stiffness. Finally, Chap. 8 deals with shooting methods for the solution for linear two-point boundary value problems. The following numerical techniques are studied in detail: simple shooting method, multiple shooting method, and the method of complementary functions and the stabilized march method for the solution to ODEs with partially separated boundary conditions.

We would like to express our thanks to those colleagues who have supported our German-Iranian joint work by their criticism and encouragement, particularly Dr. Omid Jalili (IAU, Nour Branch) for discussing most of the physical problems and Dr. Dieter Kaiser (FSU Jena) for his expert guidance on our computer programming.

It has been a pleasure working with the Springer India publication staff, in particular Mr. Shamim Ahmad and Ms. Nupoor Singh.

Finally, the authors wish to emphasize that any helpful suggestion or comment for improving this book will be warmly appreciated and can be sent by e-mail to masoud@saravi.info and martin.hermann@uni-jena.de.

Nour, Iran   Masoud Saravi
Jena, Germany   Martin Hermann

Contents

About the Authors

MARTIN HERMANN is professor of numerical mathematics at the Friedrich Schiller University (FSU) Jena (Germany). His activities and research interests are in the field of scientific computing and numerical analysis of nonlinear parameter-dependent ordinary differential equations (ODEs). He is also the founder of the Interdisciplinary Centre for Scientific Computing (1999), where scientists of different faculties at the FSU Jena work together in the fields of applied mathematics, computer sciences and applications. Since 2003, he has headed an international collaborative project with the Institute of Mathematics at the National Academy of Sciences, Kiev, Ukraine, studying, e.g., the sloshing of liquids in tanks. Since 2003, Dr. Hermann has been a curator at the Collegium Europaeum Jenense of the FSU Jena (CEJ) and the first chairman of the Friends of the CEJ. In addition to his professional activities, he volunteers in various organizations and associations. In German-speaking countries, his books *Numerical Mathematics* and *Numerical Treatment of ODEs: Initial and Boundary Value Problems* count among the standard works on numerical analysis. He has also contributed over 70 articles for refereed journals.

MASOUD SARAVI is professor of mathematics at the Islamic Azad University (IAU), Nour Branch, Iran. His research interests include the numerical solution of ODEs, partial differential equations (PDEs) and integral equations, as well as differential algebraic equations (DAEs) and spectral methods. In addition to publishing several papers with German colleagues, Dr. Saravi has published more than 15 successful titles on mathematics. The immense popularity of his books is deemed as a reflection of more than 20 years of educational experience and a result of his accessible style of writing, as well as a broad coverage of well-laid-out and easy-to-follow subjects. He is currently a board member at IAU and is working together with the Numerical Analysis Group and the Faculty of Mathematics and Computer Sciences of FSU Jena (Germany). He started off his academic studies at the UK's Dudley Technical College in the UK before receiving his first degree in mathematics and statistics from the Polytechnic of North London and his advanced degree in numerical analysis from Brunel University. After obtaining his M.Phil. in applied mathematics from the Amir Kabir University in Iran, he completed his Ph.D. in numerical analysis on solutions of ODEs and DAEs using spectral methods at the Open University in the UK.

Chapter 1
Basic Concepts of Differential Equations

1.1 Introduction

Differential equations arise in many areas of science and technology. A number of natural laws, including many from classical physics, chemistry, economics, and engineering, can be stated in the form of a differential equation.

A differential equation is a mathematical equation for an unknown function of one or several variables that relate the values of the function itself and its derivatives of various order; thus,

(i) $\quad m\dfrac{dv}{dm} + v = v^2$
$\qquad$ (ii) $\quad \dfrac{\partial^2 u}{\partial x^2} + \dfrac{\partial^2 u}{\partial y^2} + \dfrac{\partial^2 u}{\partial z^2} = 0$

(iii) $\quad \dfrac{\partial u}{\partial t} = c^2\left(\dfrac{\partial^2 u}{\partial x^2} + \dfrac{\partial^2 u}{\partial y^2}\right)$
$\qquad$ (iv) $\quad my'' + cy' + ky = 0$

(v) $\quad EIy^{(4)} = w(x)$
$\qquad$ (vi) $\quad Lq'' + Rq' + Cq = E\sin\alpha t$

(vii) $\quad y'' = k\sqrt{1 + y'^2}$
$\qquad$ (viii) $\quad x^2 y'' + xy' + (x^2 + \lambda^2)y = 0$

(ix) $\quad y'' - 2xy' + 2\lambda y = 0$
$\qquad$ (x) $\quad (1 - x^2)\,y'' - 2xy' + \lambda(\lambda + 1)y = 0$

(xi) $\quad y'' = \lambda xy$
$\qquad$ (xii) $\quad (1 - x^2)\,y'' - xy' + \lambda^2 y = 0$

are examples of differential equations. Equation (i) occurs in problems on rocket flight. Equation (ii) is the famous Laplace's equation which is found in heat, electricity, potential theory, gravitation, and aerodynamics. Equation (iii) arises in the theory of heat conduction, also in the diffusion of neutrons in an atomic pile for the production of nuclear energy. Equation (iv) has a wide range of application in the field of mechanics, in relation to simple harmonic motion, as in small oscillation

M. Hermann and M. Saravi, *A First Course in Ordinary Differential Equations: Analytical and Numerical Methods*, DOI 10.1007/978-81-322-1835-7_1, © Springer India 2014

of simple pendulum. Equation (v) is an important equation in civil engineering in the theory of bending and deflection of beams, and equation (vi) is used for RLC electric circuit. Equation (vii) is the hanging cable problem and equations (viii)–(xii) are known as Bessel's equation, Hermite's equation, Legendre's equation, Airy's equation, and Chebyshev's equation, respectively.

Of course much more different differential equations can be found in the applications, but we try to examine a number of them in this book that are in some sense representative of real-life problems. In the next chapters, we study differential equations which can be solved by analytical techniques. However, there are many equations in pure and applied physics, engineering, and other applied disciplines whose solutions cannot be represented in a closed form. Therefore, in Chaps. 7 and 8, we present modern numerical methods by which these solutions can be approximated very accurately.

1.2 Some Elementary Terminology

If a differential equation contains a single independent variable, the equation is called *ordinary differential equation* (ODE). If there are more than one independent variable, as equations (ii) and (iii), the equation is called *partial differential equation* (PDE). The dependent variable usually models the unknown quantity that has to be determined in some practical problem. For example, in the theory of heat conduction, the dependent variable u which has to be determined from equation (ii) describes a steady state of the system.

An ODE or a system of ODEs is *autonomous* if the independent variable does not appear explicitly in the equation. Autonomous equations are invariant under the translation $x \rightarrow x - a$ (see Exercise 1.5). The equation

$$y''(x) + \lambda \sin(y(x)) = 0 \qquad (1.1)$$

is an autonomous differential equation, whereas

$$y''(x) + \lambda \sin(y(x)) = \tau \cos(x)$$

is a nonautonomous equation.

The *order* of a differential equation is the order of the highest derivative present in the equation. Thus, the ODE

$$y'''(x) + 6 \left(y'(x)\right)^2 = y(x)y''(x) + e^{2y(x)}$$

has order 3. The *degree* of a differential equation is the power of the highest order derivative in the equation. The ODE

$$\left(y''(x)\right)^3 + 6y'(x) = 7\cos(x)$$

is of order 2 and degree 3.

We say a differential equation is *linear* if the dependent variable and its derivatives are of power one and it does not contain any multiplication terms of them. The ODEs

$$y''(x) + y(x) = 0 \quad \text{and} \quad y''(x) + 5y'(x) + 3y(x) = 0$$

are linear. The equation

$$y'(x) + \frac{1}{y(x)} = 0$$

is nonlinear because $1/y$ is not a first power, and Eq. (1.1) is nonlinear because $\sin(y)$ is not a first power.

A nonlinear differential equation is called *semi-linear* if its derivative of highest order has the degree one. Hence, all linear ODEs are semi-linear.

Let us look again at the set of ODEs presented in Sect. 1.1. Equations (i) and (vii) are nonlinear, with order one and two, respectively. Equations (ii), (iii), (iv), (vi), and (viii)–(xii) are linear with order two. The order of equation (v) is four, and it is also linear. All given equations are semi-linear equations. In most cases, we will deal with semi-linear equations.

Although we mentioned two partial differential equations in our examples, the main topic of this book is to consider ordinary differential equations. Therefore, whenever we talk about differential equation, we mean ordinary differential equation.

1.3 Solutions of a Differential Equation

Any ordinary differential equation of order n can be written in the general form

$$f(x, y(x), y'(x), y''(x), \ldots, y^{(n)}(x)) = 0, \quad x \in [a, b]. \tag{1.2}$$

An *explicit* solution of the ODE is an n-times continuously differentiable function $y : [a, b] \to \mathbb{R}$ which satisfies the given Eq. (1.2). Sometimes the solution can only be determined implicitly in the form $G(x, y) = 0$. In this case, we will refer to it as *implicit solution*.

An explicit solution, in which the dependent variable can be expressed in terms of elementary functions or combinations of them, is called *closed-form solution*.

Most equations which play an important role in physics and engineering have not closed-form solutions, for example, the equations (viii)–(xii).

Remark 1.1. The claim that an equation has not a closed-form solution, in general, is not correct. The fact that we know the series expansion of some elementary functions can help us to write the series solution in closed form. For example,

$$\sum_{0}^{\infty} \frac{x^{2n+\frac{1}{2}}}{(2n+1)!} = \frac{\sin x}{\sqrt{x}},$$

but it is not always possible to write easily a series expansion in closed form. In this book, we try to find the closed-form solutions in explicit or implicit form as much as possible.   □

Some ODEs can be solved easily by one or more integrations as can be seen in the next example.

Example 1.2. Suppose the acceleration of a particle is constant, say g. Find an expression for the displacement $s(t)$ at any time t.

Solution. The acceleration of a particle is given by $\dfrac{d^2s}{dt^2} = g$, which is an ODE of order two. Integrating twice with respect to t, we come to

$$s(t) = \frac{1}{2}gt^2 + c_1 t + c_2.$$

This is a closed-form solution for the given equation.   □

Let $y_1(x), \ldots, y_n(x)$ be different solutions of a linear differential equation of order n. Then the *family of solutions* in the form

$$y(x) = c_1 y_1(x) + c_2 y_2(x) + \cdots + c_n y_n(x), \quad c_i \in \mathbb{R}, \tag{1.3}$$

is also a solution of the ODE. This property, which is called *principle of superposition*, says that the linear combination of all solutions of a linear ODE is also a solution. This solution is known as the *general solution*.

Example 1.3. Show that the family of solutions given by

$$y(x) = c_1 x + c_2 x^2 + \frac{c_3}{x}$$

is a general solution of the linear ODE

$$x^3 y''' + x^2 y'' - 2xy' + 2y = 0. \tag{1.4}$$

Solution. We have

$$y' = c_1 + 2c_2 x - \frac{c_3}{x^2},$$

$$y'' = 2c_2 + 2\frac{c_3}{x^3}, \quad \text{and}$$

$$y''' = -6\frac{c_3}{x^4}.$$

By substituting y, y', y'', and y''' into the given equation, it can be seen that the given family of solutions is a general solution for this equation.   □

Often we seek a solution $y(x)$ of a first-order differential equation such that a condition $y(x_0) = y_0$ is fulfilled, where (x_0, y_0) is a prescribed initial point. Therefore, this condition is called an *initial condition*. A first-order differential equation with an initial condition is called an *initial value problem* (IVP). An IVP for the nth-order differential equation (1.2) consists of the differential equation itself and a set of n algebraic equations (the initial conditions) at a common point x_0:

$$y(x_0) = y_0, \quad y'(x_0) = y_1, \quad \ldots, \quad y^{(n-1)}(x_0) = y_{n-1},$$

where $y_0, \ldots, y_{n-1}$ are given values. The solution of an IVP has to be determined for values $x \geq x_0$, where it is assumed that the state of the system modeled by the IVP is known for $x = x_0$. Thus, the IVPs describe the *evolution* of a given system (e.g., in biology and economics). Many authors use the symbol t instead of x to indicate that the variable of an IVP is the time.

If the additional algebraic equations are related to the endpoints a and b of a given interval $[a, b]$, they are called *(two-point) boundary conditions*. A differential equation with (two-point) boundary conditions is called a *(two-point) boundary value problem* (BVP). In contrast to IVPs, the BVPs are used to model static problems (e.g., in engineering and elastomechanics). It is also possible to add boundary conditions at a, b, and some inner points of the interval $[a, b]$. In that case, we speak of *multipoint* boundary conditions and the BVP is called *multipoint* BVP.

The adding of some algebraic equations (such as initial or boundary conditions) to a given differential equation leads to a special solution (called *particular* or *specific* solution) from the corresponding family of solutions. The linear combination of all particular solutions is a general solution. For example, $y_1(x) = x$, $y_2(x) = x^2$, and $y_3(x) = 1/x$ are solutions of the ODE (1.4); thus, they are particular solutions.

We understand by a general solution of the ODE a solution which contains some parameters, and any choice of these parameters leads to a specific solution. But for nonlinear ODEs, a solution may exist that does not arise from a general solution. We call this solution a *singular solution* which is not part of the family of solutions. Since the general solution is a collection of all solutions of an ODE, in this book we do not use the term *general solution* for a nonlinear ODE. Instead we use the term *family of solutions*.

Example 1.4. Show that $y(x) = cx - c^3$ is a family of solutions for the ODE

$$(y')^3 - xy' + y = 0,$$

but $y(x) = 2\left(\frac{x}{3}\right)^{\frac{3}{2}}$ is a singular solution for this equation.

Solution. Differentiating the first given function, we obtain $y' = c$. By a simple substitution of y and y' into the ODE, it can be seen that $y(x) = cx - c^3$ is a family of solutions for this equation. But it can be easily shown that $y(x) = 2\left(\frac{x}{3}\right)^{\frac{3}{2}}$ is a singular solution, because this solution cannot be obtained from this family for any choice of the constant c. □

Example 1.5. Verify that $y_1(x) = x$ and $y_2(x) = e^x$ are particular solutions of the ODE

$$(1 - x)\, y'' + xy' - y = 0.$$

Solution. If y_1 and y_2 are solutions of the given linear equation, the linear combination of them also must be a solution. Let $y(x) = c_1 x + c_2 e^x$, then $y' = c_1 + c_2 e^x$ and $y'' = c_2 e^x$. Substituting $y(x) = c_1 x + c_2 e^x$ into the ODE leads to

$$c_2 (1 - x)\, e^x + x\, (c_1 + c_2 e^x) - (c_1 x + c_2 e^x)$$

$$= x\, (c_1 - c_1) + e^x\, ((1 - x)c_2 + x\, c_2 - c_2) = 0.$$

Hence, $y(x) = c_1 x + c_2 e^x$ is the general solution and y_1 and y_2 are particular solutions of the given differential equation. If in the general solution of this ODE we choose $c_1 = 1$ and $c_2 = 0$, we obtain $y_1(x) = x$, but if we choose $c_1 = 0$ and $c_2 = 1$, we get $y_2(x)$.   □

Example 1.6. Show that $y\, (x) = \int_a^x \dfrac{\sin t}{t}\, dt$ is a solution of

$$xy'' + y' = \cos x.$$

Solution. Differentiating the function $y(x)$ gives $y'(x) = \dfrac{\sin x}{x}$. For the second derivative, we obtain

$$y''(x) = \frac{x \cos x - \sin x}{x^2} = \frac{\cos x}{x} - \frac{\sin x}{x^2} = \frac{\cos x}{x} - \frac{y'}{x}.$$

Thus,

$$xy''(x) + y'(x) = \cos x.$$

□

Remark 1.7. Example 1.6 gives a relation between the solution of an integral and the solution of an ODE.   □

1.4 Exercises

Exercise 1.1. For the following ODEs, identify the dependent and independent variables of the ODE. State the order of them, and say whether it is linear or nonlinear.

(1) $y'' + xy' - 4x^2 y = 0$, (2) $x^5 y^3 - 4xy = 5\tan(x)$,

(3) $xyy' = 1$, (4) $|y'| + xy = 3e^x$,

(5) $y'' = \sqrt{x - y'}$, (6) $yy'' + \sin(x)y' + 12x = \cos(x)$,

(7) $y' + f(x) y = g(x)$, (8) $y'' + f(x) y' + g(x) y = h(x)$,

(9) $y' - 4|y| = x\sin(x)$, (10) $xy^4 + y' = 6\cot(xy)$.

Exercise 1.2. For the following ODEs, verify that the given function $\phi(x)$ is a solution of the differential equation:

(1) $y'' - 5y' + 4y = 0$, $\phi(x) = 3e^x$,

(2) $x^2 y'' + 4xy' + 2y = 0$, $\phi(x) = x^{-2}$,

(3) $xy'' - y' - 4x^3 y = 0$, $\phi(x) = e^{-x^2}$,

(4) $4x^2 y'' + 4xy' + (4x^2 - 1) y = 0$, $\phi(x) = \cos(x)/\sqrt{x}$,

(5) $y^{(4)} - 16y = 0$, $\phi(x) = 3\cosh(2x) - 5\sinh(2x)$,

(6) $(1 - x\cot(x))y'' - xy' + y = 0$, $\phi(x) = x - \sin(x)$,

(7) $y'' = 2x + (x^2 - y')^2$, $\phi(x) = x^3/3$,

(8) $y' = \sin(x) + y\cos(x) + y^2$, $\phi(x) = -\cos(x)$,

(9) $y = xy' + \sin^3(y')$, $\phi(x) = 5x + \sin^3(5)$,

(10) $y'' = 1 + (y')^2$, $\phi(x) = -\ln(\cos(x))$.

Exercise 1.3. For the following ODEs, find the family of solutions:

(1) $y''' = 1 + 2x - \sin(x) + 6\cos(3x)$, (2) $xy'' + y' = 1$,

(3) $\dfrac{e^x}{x^2} y' = x - 3 + \dfrac{3}{x} - \dfrac{1}{x^2}$, (4) $r\,dr = \dfrac{2\cos(\theta)}{1 - 2\sin(\theta)}\,d\theta$,

(5) $xy' + y = 3x^4$.

Exercise 1.4. Solve the given initial value problems:

(1) $y' = 2x + \cos(x)$, $y(0) = 1$,

(2) $xy' + y = \tan(x)$, $y(0) = 0$,

(3) $y'' - 6x = \sec^2(x)$, $y(0) = 1,\ y'(0) = 0$,

(4) $xy'' + y' = 24x^5$, $y(0) = y'(0) = 1$,

(5) $y''' = xe^x - \sin(x) + 1$, $y(0) = y'(0) = 0,\ y''(0) = 1$.

Exercise 1.5. Prove that if $\phi(x)$ is a solution of an autonomous ODE on $a < x < b$, then $\phi(x - x_0)$ solves the ODE on $a + x_0 < x < b + x_0$.

Exercise 1.6. An autonomous ODE can always be replaced by a nonautonomous ODE by $z = y'(x)$. The independent variable in the new differential equation is y, and the highest derivative of z with respect to y is always one less than the highest derivative of y with respect to x. Write the following autonomous ODEs in nonautonomous form:

(1) $y^{(3)} = 5yy' + 6y$,   (2) $y'y'' = y$,

(3) $4y^{(4)} = y^{(3)} - y''$,   (4) $y^{(3)} = y'' + 4y'$.

Exercise 1.7. The acceleration of a particle is given by $a(t) = -4\sin(t)$. Find expressions for its velocity $v(t)$ and displacement $s(t)$ at any time t.

Exercise 1.8. Let $N(t)$ represent the density of a population in laboratory setting at any time t. It is suggested that the rate at which $N(t)$ varies with t under the influence of the chemical is given by

$$\frac{dN}{dt} = 1 - ke^t,$$

where k is a known constant. Under the assumption that $N(0) = 200k$, find an expression for $N(t)$.

Exercise 1.9. The deflection of a 10 m long cantilever beam with constant loading is found by solving $50y^{(4)} = 3$. Find the maximum deflection of the beam. Each cantilever end requires both deflection and slope to be zero.

Exercise 1.10. For the following functions, show that they are the solutions of the given ODEs:

(a) $y(x) = x \displaystyle\int_a^x \frac{\sin(t)}{t} dt$, $xy' - y = x\sin(x)$,

(b) $y(x) = \displaystyle\int_a^x \frac{\cos(t)}{t} dt$, $xy'' + y' = -\sin(x)$,

(c) $y(x) = e^{x^2} \displaystyle\int_a^x e^{t^2} dt$, $y'' - 2xy' - 2y = 4xe^{2x^2}$,

(d) $y(x) = \displaystyle\int_x^{x^2} \frac{e^t}{t} dt$, $xy'' + y' = 4xe^{x^2} - e^x$.

Exercise 1.11. Solve the initial value problem

$$y' = e^{-3x} \cos(x), \quad y(0) = 1.$$

Verify the value of y as $x \to \infty$.

Exercise 1.12. Suppose the velocity of a particle is given by

$$v(t) = e^{-at} \left(\cos(bt) + \sin(bt) \right),$$

where a and b are positive constants. Find the displacement $s(t)$ of this particle, given that $s(0) = 0$. Determine the particle's position as $t \to \infty$.

Chapter 2
First-Order Differential Equations

2.1 Discussion of the Main Problem

Recall that the general form of an ODE of order n is

$$f(x, y, y', y'', \ldots, y^{(n)}) = 0, \quad x \in [a, b]. \tag{2.1}$$

If we choose $n = 1$, then the equation

$$f(x, y, y') = 0, \quad x \in [a, b], \tag{2.2}$$

is a first-order differential equation.

Since Eq. (2.2) is of order one, we may guess that the solution (if it exists) contains a constant c. That is, the family of solutions will be of the form

$$F(x, y, c) = 0. \tag{2.3}$$

Moreover, if we could write (2.2) explicitly as $y' = g(x, y)$, then we can search the family of solutions in the form $y = G(x, c)$. As we mentioned in Chap. 1, the constant c can be specified by a given condition.

There are in principle two different ways to treat the ODE (2.2). The first way is based on analytical manipulations that use indefinite integration and is suitable to determine an analytical solution or even the entire family of solutions. However, there is not a general method for finding an explicit expression of the solutions of (2.2), unless this equation is linear. Some nonlinear differential equations could not be solved analytically yet. The second way uses sophisticated numerical integration techniques to compute a discrete approximation of the solution of the corresponding IVPs or BVPs. In particular, these techniques can be applied in the case that it is impossible to determine an explicit expression of the solution. The first part of this book deals with analytical methods for finding solutions of (2.2). In the last two chapters, some numerical methods for solving IVPs and BVPs are presented.

M. Hermann and M. Saravi, *A First Course in Ordinary Differential Equations: Analytical and Numerical Methods*, DOI 10.1007/978-81-322-1835-7_2, © Springer India 2014

As we mentioned before, a linear equation has only one general solution, whereas nonlinear equations may also have singular solutions. Consider the IVP

$$y' = f(x, y), \quad y(x_0) = y_0. \tag{2.4}$$

This problem may have no solution or one or more than one solution. For example, the IVP

$$|y'| + |y| = 0, \quad y(0) = 1,$$

has no real solution, the IVP

$$y' + y = 0, \quad y(0) = 1,$$

has the unique solution $y = e^{-x}$, and the IVP

$$y' = y^{2/3}, \; y(0) = 0,$$

has more than one solution, namely, $y_1(x) \equiv 0$ and $y_2(x) = \frac{1}{27}x^3$. This leads to the following two fundamental questions:

- Does there exist a solution of the IVP (2.4)?
- If this solution exists, is this solution unique?

The answers to these questions are provided by the well-known *existence* and *uniqueness theorems*.

Theorem 2.1 (Peano's existence theorem). *Let $R(a, b)$ be a rectangle in the xy-plane and the point (x_0, y_0) is inside it such that*

$$R(a, b) \equiv \{(x, y) : \; |x - x_0| < a, \quad |y - y_0| < b\}.$$

If $f(x, y)$ is continuous and $|f(x, y)| < M$ at all points $(x, y) \in R$, then the IVP (2.4) has a solution $y(x)$, which is defined for all x in the interval $|x - x_0| < c$, where $c = \min\{a, b/M\}$.

Proof. See, e.g., the monograph [1].   ■

Theorem 2.2 (Uniqueness theorem of Picard and Lindelöf). *Let the assumptions of Theorem 2.1 be satisfied. If in addition $\frac{\partial f}{\partial y}(x, y)$ is continuous and bounded for all points (x, y) in R, then the IVP (2.4) has a unique solution $y(x)$, which is defined for all x in the interval $|x - x_0| < c$.*

Proof. See, e.g., the monograph [1].   ■

Moreover, we have

Theorem 2.3 (Picard iteration). *Let the assumptions of Theorem 2.2 be satisfied. Then the IVP (2.4) is equivalent to Picard's iteration method*

$$y_n(x) = y(x_0) + \int_{x_0}^{x} f(t, y_{n-1}(t))dt, \quad n = 1, 2, 3, \ldots \quad (2.5)$$

That is, the sequence $y_0, y_1, \ldots$ converges to $y(x)$.

Proof. See, e.g., the monograph [1].   ∎

We note that in this chapter our intention is not to solve a nonlinear first-order ODE by Picard's iteration method but to present Theorem 2.3 just for discussing the existence and uniqueness theorems.

Now, as before let $c \equiv \min(a, b/M)$ and consider the interval $I : |x - x_0| \le c$ and a smaller rectangle

$$T \equiv \{(x, y) : |x - x_0| \le c, \quad |y - y_0| \le b\}.$$

By the mean value theorem of differential calculus, one can prove that if $\partial f/\partial y$ is continuous in the smaller rectangle $T \subset R(a, b)$, then there exists a positive constant K such that

$$|f(x, y_1) - f(x, y_2)| \le K|y_1 - y_2| \quad \text{for all} \quad (x, y_1), (x, y_2) \in T. \quad (2.6)$$

Definition 2.4. If $f(x, y)$ satisfies inequality (2.6) for all $(x, y_1), (x, y_2) \in T$, it is said that $f(x, y)$ satisfies a uniform *Lipschitz condition* w.r.t. y in T and K is called *Lipschitz constant*.   □

Remark 2.5. We can conclude that if $\frac{\partial f}{\partial y}$ is continuous in T, then $f(x, y)$ satisfies a uniform Lipschitz condition. On the contrary, it may $f(x, y)$ satisfy a uniform Lipschitz condition, but $\frac{\partial f}{\partial y}$ is not continuous in T. For example, $f(x, y) = |x|$ satisfies a Lipschitz condition in a rectangle which contains $(0, 0)$, but $\frac{\partial f}{\partial y}$ is not continuous.   □

Now, the assumptions of Theorem 2.2 can be strengthened as follows.

Theorem 2.6. *If $f(x, y)$ is a continuous function of x and y in a closed bounded region $R(a, b)$, and satisfies a uniform Lipschitz condition, then there exists a unique solution $y(x)$ of (2.4) defined in the interval I. Furthermore, the sequence of function $\{y_n(x)\}_{n=1}^{\infty}$ defined by (2.5) converges on I to $y(x)$.*

Remark 2.7. One should put in mind that the uniform Lipschitz condition is a sufficient, but not necessary, condition for uniqueness of solutions. For example, the IVP $y' = y^{2/3}$, $y(0) = 0$ has two solutions, namely, $y \equiv 0$ and $y = \frac{1}{27}x^3$. The Lipschitz condition (2.6) is violated in any region including the line $y = 0$, because by choosing $y_1 = 0$ and y_2 arbitrarily, we obtain

$$\frac{|f(x,0) - f(x, y_2)|}{|0 - y_2|} = \frac{1}{\sqrt[3]{y}}$$

which can be made large as much as we please by choosing y_2 sufficiently small.

□

The continuity of $f(x, y)$ alone is not sufficient for the convergence of Picard's iteration method, as indicated by the following example.

Example 2.8. Let $f(x, y(x))$ be defined on the rectangle $R \equiv [-1, 1] \times [-1, 1]$ as

$$f(x, y(x)) = \begin{cases} 0, & \text{if } x = 0 \\ \dfrac{2y}{x}, & \text{if } |x| \le 1, x \ne 0, \ |y| \le x^2 \\ 2x, & \text{if } |x| \le 1, \ y > x^2 \\ -2x, & \text{if } |x| \le 1, \ y < -x^2 \end{cases}$$

Prove that for all points $(x, y) \in R$ the function $f(x, y(x))$ is bounded but does not satisfy a uniform Lipschitz condition. Then conclude that the successive approximations generated by Picard's iteration method do not converge to the solution $y(x)$ of the IVP $y' = f(x, y)$, $y(0) = 0$.

Solution. We have

$$|f| = \begin{cases} 0, & \text{if } x = 0 \\ \left|\dfrac{2y}{x}\right| = \dfrac{2x^2}{|x|} = 2|x|, & \text{if } 0 < x \le 1, \ |y| \le x^2 \\ |2x| = 2, & \text{if } 0 < x \le 1, \ y > x^2 \\ |-2x| = 2, & \text{if } 0 < x \le 1, \ y < -x^2 \end{cases}$$

Hence, $|f| \le 2$ because $0 < x \le 1$, i.e., f is bounded and continuous. Furthermore, for the IVP $y' = f(x, y)$, $y(0) = 0$, and $0 \le x \le 1$, the successive approximations are $y_1 = 0$, $y_2 = x^2$, and $y = -x^2$. Thus, the successive approximations do not converge to the solution $y(x)$ of the IVP.

□

We have decided not to provide the proofs of Theorems 2.1–2.3 and 2.6, but we included some remarks and examples which may be helpful for a better understanding of these theorems.

2.2 Analytical Solution of First-Order ODEs

In the previous chapter, we solved some simple ODEs. Now we turn our attention to more general cases. As we pointed out, it may be difficult to find closed-form solutions whenever we deal with nonlinear ODEs. We are going to introduce some analytical methods, known as elementary methods, which are based on integration.

Let us write (2.2) as

$$M(x, y)\, dx + N(x, y)\, dy = 0. \tag{2.7}$$

There are special cases for which this equation can be solved analytically. The first one is that of a *separable* equation.

Definition 2.9. The ODE (2.7) is called separable if it can be transformed into an equation

$$M(x)dx + N(y)dy = 0. \tag{2.8}$$

□

Then we say the variables are separable, and the solution can be obtained by integrating both sides with respect to x. Thus,

$$\int M(x)dx + \int N(y(x))\frac{dy}{dx}\, dx = c,$$

or for simplicity one can write

$$\int M(x)dx + \int N(y)dy = c.$$

Example 2.10. Solve the IVP

$$e^x y' + x\left(1 + y^2\right) = 0, \quad y(-1) = 0.$$

Solution. The given ODE can be written as

$$\frac{dy}{1 + y^2} = -xe^{-x}dx.$$

Integrating both sides of the equation yields

$$\tan^{-1}(y) = (x + 1)\, e^{-x} + c.$$

Using the initial condition $y(-1) = 0$, we obtain $c = 0$. Hence, $\tan^{-1}(y) = (x + 1)\, e^{-x}$, and

$$y(x) = \tan((x + 1)e^{-x}).$$

□

Example 2.11 (Decay problem). This problem occurs in various fields like the decay of radioactive elements in physics and chemistry. It has been found that a radioactive substance decomposes at a rate, called the *decay rate*, proportional to the present amount. That is, if m is the mass of the substance at the present time t, then the rate of change per unit mass, $\left(\frac{dm}{dt}\right)/m$, is a constant. If at the time $t = 0$ the amount of the substance is m_0, what can be said about the amount available at any time $t > 0$?

If *half-life* denotes the time required for a quantity to fall to half its value as measured at the beginning of the time period, find the decay rate of a substance whose half-life is 50 years.

Solution. We use the differential equation

$$\frac{dm}{dt} = -k\,m, \quad k > 0.$$

We rewrite it as

$$\frac{dm}{m} = -k\,dt,$$

hence the variables are separable. By integration, we obtain

$$\ln(m) = -k\,t + \ln(c).$$

Thus,

$$\ln\left(\frac{m}{c}\right) = -k\,t,$$

hence $m = c\,e^{-kt}$. The initial condition $m(0) = m_0$ leads to $c = m_0$. It follows

$$m(t) = m_0\,e^{-kt}.$$

Note, for simplicity we have used $\ln(c)$ instead of c as constant of integration.

For the second part of this example, let H be the corresponding half-life. Then

$$\frac{m_0}{2} = m_0\,e^{-kH} \rightarrow \frac{1}{2} = e^{-kH} \rightarrow \ln(2) = 50\,k,$$

and we obtain

$$k = \frac{\ln(2)}{50} = \frac{0.693147}{50} = 0.013863.$$

□

Example 2.12 (Newton's law of cooling). Consider a heated object placed in an environment whose surrounding temperature is lower than that of the object. Newton's law of cooling states that the rate of change of the temperature of the object is proportional to difference between the temperature of the object and that of its surrounding. Let $T(t)$ be the temperature of the object at any time t and T_0 be the (constant) temperature of the surrounding, and then the equation which describes this law may be written as

$$T'(t) = k \ (T(t) - T_0),$$

where k is a constant. Show that $T(t) - T_0$ and k are negative. Solve this equation using the initial condition $T(0) = 1 + T_0$.

Solution. The given ODE can be reformulated as

$$\frac{dT}{T(t) - T_0} = k \, dt.$$

Integrating both sides, we obtain

$$T(t) = T_0 + c \, e^{kt}.$$

The initial condition $T(0) = 1 + T_0$ gives

$$T(t) = T_0 + e^{kt}.$$

$\square$

Another special case of (2.7) are *homogeneous* ODEs.

Definition 2.13. The function $f(x, y)$ is homogeneous of degree k if it holds $f(tx, ty) = t^k f(x, y)$, for some nonzero constant k. If the right-hand side $f(x, y)$ of the ODE (2.4) is homogeneous, then the differential equation is called homogeneous. $\square$

If in (2.7), $M(x, y)$ and $N(x, y)$ have the same degree of homogeneity, or in (2.4) the function $f(x, y)$ has the degree zero, then the substitution $z = y/x$ will transform the given ODE into a separable form. Let $y = xz$, then $y' = z + xz'$. Substituting this into the ODE gives $z + xz' = f(z)$. Now separating the variables, we obtain

$$\frac{z'}{f(z) - z} = \frac{1}{x}.$$

Example 2.14. Solve the ODE $y' = \dfrac{x^2 + 2y^2}{2xy}$.

Solution. The degree of homogeneity of the right-hand side

$$f(x, y) = \frac{x^2 + 2y^2}{2xy}$$

Fig. 2.1 Geometry used in
Example 2.15

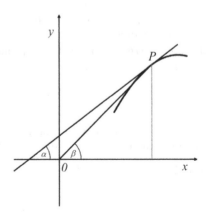

is zero. Hence, the substitution $y = xz$ and $y' = z + xz'$ leads to the following
separable form $2zdz = dx/x$. Thus, $z^2 = \ln|x| + c$. Since $z = y/x$, we get
$y^2 = x^2 (\ln|x| + c)$.   □

Example 2.15. Find the curve in the xy-plane such that the angle of the tangent line
at any of its points P is three times as big as the angle of inclination of OP (see
Fig. 2.1).

Solution. We have $\alpha = 3\beta$, i.e., $\tan(\alpha) = \tan(3\beta)$. Let us write

$$\tan(\alpha) = \tan(\beta + 2\beta) = \frac{\tan(\beta) + \tan(2\beta)}{1 - \tan(\beta)\tan(2\beta)} = \frac{3\tan(\beta) - \tan^3(\beta)}{1 - 3\tan^2(\beta)}.$$

A look at Fig. 2.1 shows that $\tan(\alpha) = y'$ and $\tan(\beta) = y/x$; hence,

$$y' = \frac{3\left(\frac{y}{x}\right) - \left(\frac{y}{x}\right)^3}{1 - 3\left(\frac{y}{x}\right)^2} = \frac{3yx^2 - y^3}{x^3 - 3xy^2}.$$

Thus, $f(x, y) = \dfrac{3yx^2 - y^3}{x^3 - 3xy^2}$ has the degree of homogeneity zero.

The substitution $y = xz$ and $y' = z + xz'$ leads to

$$\frac{(1 - 3z^2)}{z(1 + z^2)}dz = \frac{2}{x}dx,$$

whose solution is

$$\frac{z}{(1 + z^2)^2} = cx^2.$$

With $z = y/x$, the solution can be written in implicit form as

$$xy - c(x^2 + y^2)^2 = 0.$$

$\square$

Remark 2.16. It is possible that a given ODE may not be homogeneous, but the ansatz $y = xz$ changes it to a separable equation. $\square$

Example 2.17. Solve the IVP

$$y' = \frac{x^3 e^x + y}{x}, \quad y(1) = 0.$$

Solution. This is not a homogeneous ODE, but the substitution $y = xz$ gives $z' = xe^x$, which is a separable equation. By integration, we obtain $z = (x - 1) e^x + c$. Setting $z = y/x$ leads to $y = x[(x - 1) e^x + c]$. The initial condition $y(1) = 0$ determines the solution $y(x) = x(x - 1) e^x$. $\square$

As a next special case of (2.7), we will now discuss so-called *exact* ODEs.

Definition 2.18. The total differential of a function $z(x, y)$ is given by

$$dz = z_x dx + z_y dy. \tag{2.9}$$

$\square$

The connection between (2.7) and (2.9) may lead to a solution strategy for the ODE (2.7).

Definition 2.19. We say that the ODE (2.7) is exact if

$$M(x, y) dx + N(x, y) dy$$

is the total differential of a function $z(x, y)$ so that the ODE may be written as $dz = 0$. $\square$

Theorem 2.20. *If $M(x, y)$ and $N(x, y)$ have continuous first-order partial derivatives over some domain D, then (2.7) is exact if and only if $M_y = N_x$.*

Proof. (i) Let

$$M(x, y)dx + N(x, y)dy = 0 \tag{2.10}$$

be exact. Then there must be a function $z(x, y)$ such that

$$\frac{\partial z}{\partial x} = M, \quad \frac{\partial z}{\partial y} = N. \tag{2.11}$$

Because of the assumptions on M and N, we can say that

$$\frac{\partial}{\partial y}\left(\frac{\partial z}{\partial x}\right) \quad \text{and} \quad \frac{\partial}{\partial x}\left(\frac{\partial z}{\partial y}\right)$$

exist and

$$\frac{\partial}{\partial y}\left(\frac{\partial z}{\partial x}\right) = \frac{\partial}{\partial x}\left(\frac{\partial z}{\partial y}\right). \tag{2.12}$$

Substituting (2.11) into (2.12), we get $M_y = N_x$.

(ii) When the ODE is exact, then we have $z_x = M$ and $z_y = N$. Define $z(x, y)$ by

$$z(x, y) \equiv \int_x M(x, y)dx + g(y).$$

From this, we obtain

$$z_y = \frac{\partial}{\partial y}\int_x M(x, y)dx + g'(y)$$

and this is equal to $N(x, y)$. Therefore,

$$g'(y) = N(x, y) - \frac{\partial}{\partial y}\int_x M(x, y)dx.$$

We can find $g(y)$ whenever $g'(y)$ is independent from x, and this is true if the derivative w.r.t. x of the right-hand side of the ODE is zero. Performing this differentiation leads to

$$\frac{\partial}{\partial x}\left[N(x, y) - \frac{\partial}{\partial y}\int_x M(x, y)dx\right] = \frac{\partial N}{\partial x} - \frac{\partial}{\partial x}\frac{\partial}{\partial y}\int_x M(x, y)dx$$

$$= \frac{\partial N}{\partial x} - \frac{\partial M}{\partial y}.$$

Since the ODE is exact, the last expression will be zero.   □

Example 2.21. Find the family of solutions of the following ODE:

$$y' = \frac{e^x - xy^2}{x^2y + \sin(y)}.$$

Solution. Obviously, the variables are not separable. We write the given ODE in the form

$$\left(xy^2 - e^x\right)dx + \left(x^2y + \sin(y)\right)dy = 0.$$

Let us set $M(x, y) \equiv xy^2 - e^x$ and $N(x, y) \equiv x^2y + \sin(y)$. It can be easily seen that $M_y = N_x = 2xy$. Hence, the ODE is exact and we have the relation $z_x = M(x, y) = xy^2 - e^x$. Integration w.r.t. x gives

$$z(x, y) = \frac{x^2y^2}{2} - e^x + g(y).$$

Differentiation w.r.t. y leads to

$$z_y(x, y) = x^2y + g'(y).$$

Now, since $z_y(x, y) = N(x, y) = x^2y + \sin(y)$, we have

$$g'(y) = \sin(y) \quad \rightarrow \quad g(y) = -\cos(y).$$

Therefore, the family of solutions will be

$$\frac{x^2y^2}{2} - e^x - \cos(y) = c.$$

$\square$

Remark 2.22. Another technique to solve exact ODEs is as follows. Since the ODE is exact, we can write it in the form

$$A(x)\,dx + B(y)\,dy + dF(x, y) = 0,$$

where

$$dF(x, y) \equiv C(x, y)dx + D(x, y)dy.$$

Integrating both sides, we obtain

$$\int A(x)dx + \int B(y)dy + F(x, y) = c.$$

For Example 2.21, one can write

$$(-e^x dx) + (\sin(y)dy) + \left(xy^2 dx + x^2 y dy\right) = 0.$$

Integrating both sides, we obtain

$$-e^x - \cos(y) + \frac{x^2y^2}{2} = c.$$

$\square$

Example 2.23. Find the solution of the IVP

$$ye^x dx + (e^x - \sin(y)) \, dy = 0, \quad y(0) = 0.$$

Solution. We set $M_y = e^x$ and $N_x = e^x$. Since it holds $M_y = N_x$, the given ODE is exact. To solve this equation, we write it in the form

$$(-\sin(y)) \, dy + (ye^x dx + e^x dy) = 0,$$

and integrate. We obtain

$$\int (-\sin(y)) \, dy + \int (ye^x dx + e^x dy) = c.$$

It can be easily shown that the solution of this equation is given by

$$\cos(y) + ye^x = c.$$

The initial condition $y(0) = 0$ determines the constant of integration c as $c = 1$. Hence, $\cos(y) + ye^x = 1$ is the solution of the given IVP. $\qquad\square$

Let us now discuss the solution of (2.7) under the assumption that the ODE is not exact, i.e., $M_y \neq N_x$.

To find a solution in this case may be difficult or impossible, even there exists a solution. But sometimes (2.7) may be made exact by multiplying it by a function $F(x, y) (\neq 0)$, called *integrating factor*, that usually is not unique. If such an integrating factor exists, then the modified differential equation

$$F(x, y) M(x, y) + F(x, y) N(x, y) = 0$$

will be exact. Therefore, $(FM)_y = (FN)_x$, i.e.,

$$MF_y + FM_y = NF_x + FN_x.$$

We can rewrite this equation as

$$\frac{NF_x - MF_y}{F} = M_y - N_x. \tag{2.13}$$

This is a first-order partial differential equation that may have more than one solution. But the determination of these solutions will be more difficult than to solve the original equation (2.7). Fortunately, if we restrict the integrating factor to be a single variable function in terms of x or y, we can find solutions. Suppose $F = F(x)$, then $F_y = 0$ and (2.13) reduces to

$$\frac{F_x}{F} = \frac{M_y - N_x}{N}.$$

It is not difficult to show that

$$F = e^{\int \frac{M_y - N_x}{N} dx}.$$ (2.14)

On the other hand, if $F = F(y)$, then $F_x = 0$, and we can easily show that

$$F = e^{-\int \frac{M_y - N_x}{M} dy}.$$ (2.15)

Thus, if $(M_y - N_x)/N$ is a function of x, then $F = F(x)$ or if $(M_y - N_x)/M$ is a function of y, then $F = F(y)$.

Example 2.24. Determine a particular solution of the IVP

$$\sin(y)\cos(x)dx + \cos(y)[\sin(y) - \sin(x)]dy = 0, \quad y(0) = \frac{\pi}{2}.$$

Solution. We have $M_y = \cos(x)\cos(y)$ and $N_x = -\cos(x)\cos(y)$. Hence, the equation is not exact, but it holds

$$\frac{M_y - N_x}{M} = 2\cot(y) \equiv g(y).$$

Thus,

$$F(y) = e^{-\int g(y)dy} = e^{\int -2\cot(y)dy} = e^{-2\ln(\sin(y))} = \frac{1}{\sin^2(y)}.$$

Multiplying the differential equation by $1/\sin^2(y)$, we get

$$\frac{\cos(x)}{\sin(y)}dx + \frac{\cos(y)}{\sin^2(y)}[\sin(y) - \sin(x)]dy = 0.$$

This equation is exact. Let us rewrite it as

$$\left(\frac{\cos(y)}{\sin(y)}dy\right) + \left(\frac{\cos(x)}{\sin(y)}dx - \frac{\cos(y)\sin(x)}{\sin^2(y)}dy\right) = 0,$$

and integrate both sides. The solution of the resulting equation is

$$\ln(\sin(y)) + \frac{\sin(x)}{\sin(y)} = c.$$

To determine the particular solution satisfying the prescribed initial condition, we substitute $x = 0$ and $y = \pi/2$ into this family of solutions, obtaining $c = 0$. Hence, the solution is given by

$$\ln(\sin(y)) + \frac{\sin(x)}{\sin(y)} = 0.$$

□

Example 2.25. Solve $2\left(x^2 y^2 + y + x^4\right) dx + x\left(1 + x^2 y\right) dy = 0$.

Solution. This equation is not exact because

$$M_y = 4x^2 y + 2 \quad \text{and} \quad N_x = 1 + 3x^2 y,$$

but we have $(M_y - N_x)/N = 1/x = g(x)$. Hence,

$$F(x) = e^{\int g(x)dx} = e^{\int (1/x)dx} = x.$$

Multiplying the given equation by this integrating factor, we obtain

$$2x\left(x^2 y^2 + y + x^4\right) dx + x^2\left(1 + x^2 y\right) dy = 0,$$

which is an exact equation. Now we rewrite this equation as

$$(2x^5 dx) + \left(2x\left(x^2 y^2 + y\right) dx + x^2\left(1 + x^2 y\right) dy\right) = 0.$$

Its solution can be found by integration and we obtain

$$2x^6 + 3x^4 y^2 + 6x^2 y = c.$$

□

As a next special case of (2.7), we consider now *linear* ODEs. The general form of a linear ODE is given by

$$y' + p(x) y = q(x). \tag{2.16}$$

When $q(x) = 0$, the equation is reduced to a separable equation. Assume $q(x) \neq 0$ and consider (2.16). This equation is not exact, but it can easily be shown that it has an integrating factor given by

$$F(x) = e^{\int p(x)dx}.$$

Multiplying (2.16) by this factor, we get

$$e^{\int p(x)dx}(y' + p(x) y) = e^{\int p(x)dx} q(x). \tag{2.17}$$

This equation is exact. Integrating both sides, we obtain

$$e^{\int p(x)dx} y = \int e^{\int p(x)dx} q(x)\, dx + c.$$

Hence,

$$y = e^{-\int p(x)dx} \int e^{\int p(x)dx} q(x)\, dx + c e^{-\int p(x)dx}, \tag{2.18}$$

which is the general solution of (2.16).

Remark 2.26. The integration of the left-hand side of (2.17) will always be the same given by

$$\text{function } y(x) \times \text{integrating factor } F(x).$$

$\square$

Example 2.27. Solve the ODE $xy' - 2y = x^3 \cos(x), \quad x > 0$.

Solution. We write

$$y' - \frac{2}{x} y = x^2 \cos(x). \tag{2.19}$$

Hence, the integrating factor will be given by

$$F(x) = e^{\int p(x)dx} = e^{\int (-\frac{2}{x})dx} = e^{-2\ln(x)} = x^{-2}.$$

We multiply both sides of (2.19) by x^{-2}. We obtain

$$x^{-2}\left(y' - \frac{2}{x} y\right) = \cos(x).$$

By simple integration, we come to the result

$$x^{-2} y = \sin(x) + c \quad \rightarrow \quad y(x) = x^2(\sin(x) + c).$$

$\square$

Linear first-order ODEs have various applications in physics and engineering. To illustrate this, let us consider some standard examples.

Example 2.28 (Electric circuit). Consider an electric circuit (see Fig. 2.2) with negligible capacitance but containing a resistor R, an inductor L, and an electromotive force $E(t)$.

Find the current in this LR-circuit. If we assume $E(t) = E_0$ is a constant, describe the situation as $t \to \infty$.

Fig. 2.2 Electric circuit
considered in Example 2.28

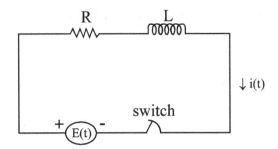

Solution. According to Kirchhoff's voltage law, the sum of the voltage drop across
the inductor $L(di/dt)$ and voltage drop across the resistor Ri is the same as the
impressed voltage $E(t)$ on the circuit. Hence, we get the ODE

$$L\frac{di}{dt} + Ri = E(t).$$

It is not difficult to show that the general solution of this equation is

$$i(t) = \frac{e^{-(\frac{R}{L})t}}{L} \int e^{(\frac{R}{L})t} E(t)\, dt + c\, e^{-(\frac{R}{L})t}.$$

If we substitute $E(t) = E_0$, then we obtain

$$i(t) = \frac{E_0}{R} + c\, e^{-(\frac{R}{L})t}.$$

As $t \to \infty$, the second term vanishes. This term is called the *transient term* and
the remaining term is known as the *steady-state current*.   □

Example 2.29 (Mixing problem).   A tank contains V_0 gallons brine in which A (lb)
of salt are dissolved. Another solution containing B (lb) of salt per gallon flows into
the tank at the rate of C gal/min. If the well-stirred solution flows out at the rate of
D gal/min, find the amount of salt at any time t.

Solution. Let $S(t)$ be the amount of salt at any time. The rate at which the salt flows
into the tank is BC lb/min and the rate at which the salt flows out is

$$\left(\frac{D}{V_0 + Ct - Dt}\right) S(t),$$

because the volume of brine at any time t is the initial volume V_0 plus the value of
brine added Ct minus the volume of brine removed Dt. Since

$$\frac{dS}{dt} = (\text{rate } S \text{ flows in}) - (\text{rate } S \text{ flows out}),$$

we get

$$\frac{dS}{dt} + \left(\frac{D}{V_0 + (C - D)t}\right) S = BC.$$

It is not difficult to show that the integrating factor $F(t)$ is

$$F(t) = [V_0 + (C - D)t]^{\frac{D}{C-D}}.$$

Multiplying the ODE by $F(t)$ and subsequent integration yields the solution

$$S(t) = B[V_0 + (C - D)t] + c[V_0 + (C - D)t]^{\frac{D}{D-C}}.$$

□

We now come to the so-called *reducible ODEs*.

Some of the nonlinear ODEs may be reduced to linear or separable equations by forming the dependent variable $z = g(y)$. The most common examples of such ODEs are the Bernoulli and the Riccati equations. In the rest of this section, we consider these two equations.

(a) *Bernoulli equation*

The general form of a Bernoulli equation is given by

$$y' + p(x)y = q(x)y^n, \tag{2.20}$$

where n is a real number. Applications of this equation are in the study of population dynamics and in hydrodynamic stability. When $n = 0$ or 1, this equation reduces to a linear equation; thus, we assume $n \neq 0$ or 1.

If we set $z \equiv y^{1-n}$, then $z' = (1-n)y'y^{-n}$. That is, $y' = \frac{1}{1-n}y^n z'$. Substituting this into (2.20) yields

$$z' + (1 - n)p(x)z = (1 - n)q(x), \tag{2.21}$$

which is a linear first-order ODE.

Example 2.30. Solve the ODE $xy' + 2y = x\sqrt{y}$.

Solution. We write the given ODE in the form

$$y' + \frac{2}{x}y = y^{1/2}.$$

This is a Bernoulli equation with $n = 1/2$. Hence, $z = y^{1-1/2} = y^{1/2}$. Substituting this into the Bernoulli equation, we obtain

$$z' + \frac{1}{x}z = \frac{1}{2}.$$

It is not difficult to show that the general solution of this equation is

$$xz = \frac{x^2}{4} + c.$$

Since $z = y^{1/2}$, we have

$$xy^{1/2} = \frac{x^2}{4} + c \quad \text{or} \quad y(x) = \left(\frac{x}{4} + \frac{c}{x}\right)^2.$$

□

Example 2.31. Let us consider the two-dimensional projectile motion under the influence of gravity and air resistance. Using the assumption that the air resistance is proportional to the square of the velocity, the model equation relating the speed $s(t)$ of the projectile with the angle of inclination $\theta(t)$ will be given by

$$\frac{ds}{d\theta} - \tan(\theta)s = b \sec(\theta)s^3,$$

where b is a constant which measures the amount of air resistance. Solve this equation.

Solution. The ODE is a Bernoulli equation with $n = 3$. Let $z \equiv s^{1-3} = s^{-2}$. Hence, $z' = -2s's^{-3}$. Substituting $s' = -(z'/2)s^3$ into the ODE gives

$$\frac{dz}{d\theta} + 2\tan(\theta)z = -2b \sec(\theta).$$

This is a linear ODE and one can show that its general solution is

$$z(\theta) = \frac{c - b\left(\sec(\theta)\tan(\theta) + \ln(\sec(\theta) + \tan(\theta))\right)}{\sec^2(\theta)}.$$

Since $z = s^{-2}$, we can write

$$s(\theta) = \frac{\sec(\theta)}{\sqrt{c - b(\sec(\theta)\tan(\theta) + \ln(\sec(\theta) + \tan(\theta)))}}.$$

□

(b) *Riccati equation*

The general form of a Riccati equation is given by

$$y' + p(x)y = q(x)y^2 + r(x). \tag{2.22}$$

Obviously, if $r(x) = 0$, we will have a Bernoulli equation with $n = 2$. Therefore, we assume $r(x) \neq 0$. Equation (2.22) cannot be solved by elementary methods, but if a particular solution y_1 is known, then the solution of this ODE can be obtained through the substitution $y = y_1 + 1/z$, where z is the general solution of

$$z' + (2q(x)y_1 - p(x))z = -q(x).$$

Example 2.32. First, show that $y = \cos(x)$ is a particular solution of the ODE

$$(1 - \sin(x)\cos(x))y' - y + \cos(x)y^2 = -\sin(x),$$

then solve this equation.

Solution. We have $y = \cos(x)$; hence, $y' = -\sin(x)$. Substituting this into the ODE leads to

$$(1 - \sin(x)\cos(x))(-\sin(x)) - \cos(x) + \cos(x)\cos^2(x)$$

$$= -\sin(x) + \sin^2(x)\cos(x) - \cos(x) + \cos(x)\left(1 - \sin^2(x)\right)$$

$$= -\sin(x).$$

Thus, $y = \cos(x)$ is a particular solution for this equation.

Now, let $y = \cos(x) + 1/z$ then $y' = -\sin(x) - z'/z^2$. Substituting this into the ODE, we get

$$\left(1 - \sin(x)\cos(x)\right)\left(-\sin(x) - \frac{z'}{z^2}\right) - \left(\cos(x) + \frac{1}{z}\right) + \cos(x)\left(\cos(x) + \frac{1}{z}\right)^2$$

$$= -\sin(x).$$

By simplifying this expression, we obtain

$$\left(1 - \sin(x)\cos(x)\right)z' + (1 - 2\cos^2(x))z = \cos(x).$$

Using $(1 - \sin(x)\cos(x))' = 1 - 2\cos^2(x)$ and integrating it, we get

$$z = \frac{\sin(x) + c}{1 - \sin(x)\cos(x)}.$$

We also have $y = \cos(x) + 1/z$, that is, $z = 1/(y - \cos(x))$. Therefore,

$$y(x) = \frac{c\cos(x) - 1}{c - \sin(x)}.$$

$\square$

Remark 2.33. Another choice for $y(x)$ is to write $y(x) = y_1(x) + z(x)$, where $z(x)$ is the family of solutions of the equation

$$z' + (p(x) - 2q(x)y_1) = q(x)z^2.$$

□

Example 2.34. Solve the ODE $y' + 2xy = y^2 + x^2 + 1$, if its particular solution is given by $y_1(x) = x$.

Solution. Let $y = x + z$. Then, $y' = 1 + z'$. Substituting this into the ODE, we obtain $z' = z^2$. Now, it is not difficult to show that the family of solutions of this equation will be

$$z = \frac{1}{c - x}.$$

That is, the solution is $y(x) = x + \dfrac{1}{c - x}$.   □

At the end of this section, we will consider the so-called method of change of variables.

Often a change of variables may be useful to solve an ODE. This choice can be $z = g(x)$ or $z = h(y)$. Although this choice of the variables seems to be complicated, in many cases the presence of the variables helps us to apply this technique.

Example 2.35. Solve the IVP

$$y' - \frac{y \ln(y)}{x + 1} = (x + 1)y, \quad y(0) = 1.$$

Solution. Because of the presence of the function $\ln(y)$ in the ODE, one can guess $z = \ln(y)$ or $y = e^z$. Hence, $y' = e^z z'$. Substituting this into the ODE and simplifying, we get

$$z' - \frac{z}{x + 1} = x + 1.$$

This is a linear equation and it can be easily shown that its particular solution is

$$z = x(x + 1).$$

Since $y = e^z$, we obtain

$$y(x) = e^{x(x+1)}.$$

□

Example 2.36. Solve the ODE

$$xyy' = 6\left(3y^2 + 2\right) + x\left(3y^2 + 2\right)^2.$$

Solution. The presence of $\left(3y^2 + 2\right)$ encourages us to choose $z = 3y^2 + 2$. Hence, $z' = 6yy'$. If we substitute this into the ODE, after simplification, we obtain

$$z' = \frac{z}{x} + \frac{1}{6}z^2,$$

which is a Bernoulli equation with $n = 2$. Let $t = z^{1-2} = z^{-1}$. Then we get

$$xt' + t = -\frac{1}{6}x.$$

The general solution of this linear equation can easily be determined as

$$t = \frac{c}{x} - \frac{x}{12}.$$

Since

$$t = z^{-1} = (3y^2 + 2)^{-1},$$

the required solution is given by

$$(3y^2 + 2)^{-1} = \frac{c}{x} - \frac{x}{12}.$$

$\square$

2.3 High-Degree First-Order ODEs

So far we have considered first-order ODEs with degree one. In this section, we discuss, briefly, equations with a higher degree.

The general form of this type of equations can be expressed as

$$(y')^n + p_1(x, y)\left(y'\right)^{n-1} + \cdots + p_{n-1}(x, y)\, y' + p_n(x, y) = 0. \qquad (2.23)$$

Unfortunately, solving (2.23) is very difficult and in most cases impossible even if there exist a solution. However, there are some special cases of (2.23) which can be solved analytically. Here we will discuss two of these special cases.

(a) *Equations solvable for y'*

Suppose (2.23) can be factorized as

$$[y' - f_1(x, y)] [y' - f_2(x, y)] \cdots [y' - f_n(x, y)] = 0.$$

Then, we set each factor equal to zero, i.e.,

$$y' - f_i(x, y) = 0, \quad i = 1, 2, \ldots, n,$$

and determine the solutions

$$\Phi_i(x, y, c_i) = 0, \quad i = 1, 2, \ldots, n,$$

of these factors. Now, the family of solutions of (2.23) can be written in the form

$$\Phi_1(x, y, c_1) \, \Phi_2(x, y, c_2) \cdots \Phi_n(x, y, c_n) = 0.$$

Example 2.37. Solve the ODE

$$x^2 (y')^2 + xy' - y^2 - y = 0.$$

Solution. In terms of y', this equation can be compared with an algebraic equation of degree two. Hence, we get

$$(xy' - y)(xy' + y + 1) = 0.$$

The solution of the first factor is $y - c_1 x = 0$ and the solution of the second one is $x(y + 1) - c_2 = 0$. Thus,

$$(y - c_1 x)(x(y + 1) - c_2) = 0$$

will be the solution. $\square$

(b) *Equations solvable for x or y*

Suppose we can write (2.23) in the form

$$y = f(x, y'). \tag{2.24}$$

Differentiating (2.24) w.r.t. x gives

$$y' = F(x, p, p'), \tag{2.25}$$

where $y' = p$. The ODE (2.25) depends on the two variables x and p. We write its solution in the form

$$\Phi(x, p, c) = 0. \tag{2.26}$$

The elimination of $p = y'$ from (2.24) and (2.26) gives the desired solution. Similar operations can be realized when the independent variable is missing.

These methods can be helpful for solving (2.23), but when $n \geq 3$, we encounter difficulties to factorize (2.23) or eliminate y'. To see this, let us consider the equation

$$y = 3xy' - (y')^3.$$

Differentiating it w.r.t. x yields

$$y' = 3y' + 3xy'' - 3y''(y')^2, \quad \text{i.e.} \quad 3(x - p^2)p' = -2p.$$

If we rewrite this equation in the form

$$2p\,dx + 3(x - p^2)\,dp = 0,$$

then it is not difficult to show that it has the integrating factor $F = \sqrt{p}$. Multiplying this equation by F and integrating term by term, we obtain

$$p\sqrt{p}(x - p^2) = c, \quad \text{i.e.} \quad y'\sqrt{y'}(x - (y')^2) = c.$$

The next step would be the elimination of y' from this and the main equation. But, as we observe, this elimination is nearly impossible.

Closing this section, let us consider two special ODEs where this elimination process is indeed possible.

(i) *Clairaut equation*

The general form of a Clairaut equation is given by

$$y = xy' + f(y'). \tag{2.27}$$

Differentiating this ODE w.r.t. x gives

$$[x + f'(p)]\frac{dp}{dx} = 0,$$

where $y' = p$. We can have

$$\frac{dp}{dx} = 0 \quad \text{or} \quad x + f'(p) = 0.$$

If $dp/dx = 0$, then $p = y' = c$. Now using $y' = c$ in (2.27), we obtain

$$y(x) = cx + f(c). \tag{2.28}$$

This is the family of solutions of the Clairaut equation. One can eliminate y' from (2.27) and $x + f'(y') = 0$ to obtain a singular solution because it does not contain c.

Example 2.38. Solve the ODE $y = xy' - 2(y')^3$, and find its singular solution.

Solution. This is a Clairaut equation; hence, its family of solutions will be $y = cx - 2c^3$. We have $x + f'(y') = x - 6(y')^2 = 0$. Now, if we eliminate y' from this and the given equation, we obtain

$$y(x) = 4 \left(\frac{x}{6}\right)^{3/2},$$

which is a singular solution.   □

(ii) *Lagrange equation*

A more general form of a Clairaut equation is given by

$$y = f(y') + xg(y') \tag{2.29}$$

and is known as Lagrange equation. If we differentiate the ODE w.r.t. x and set $y' = p$, we obtain

$$\frac{dx}{dp} - \frac{g'(p)}{p - g(p)} x = \frac{f'(p)}{p - g(p)}.$$

Obviously, this is a linear equation and can be solved by the formula given in (2.18).

Example 2.39. Solve the ODE

$$y = x(y')^2 - (y')^2.$$

Solution. This is a Lagrange equation with $f(y') = -g(y') = (y')^2$. Therefore,

$$\frac{dx}{dp} - \frac{2p}{p - p^2} x = \frac{-2p}{p - p^2}.$$

That is,

$$\frac{dx}{dp} + \frac{2}{p - 1} x = \frac{-2}{p - 1}.$$

The general solution of this equation can be found by (2.18). We obtain

$$x = 1 + \frac{c}{(p - 1)^2} = 1 + \frac{c}{(y' - 1)^2}.$$

By elimination of y' from this and the given equation, we get

$$y(x) = x + c + 2\sqrt{\frac{c}{x-1}}.$$

□

What makes the solution of Lagrange equations tedious is the elimination of y'. But this is the same difficulty which we have already encountered in the case (b).

2.4 Family of Curves, Orthogonal Trajectories

Let us recall Eq. (2.3)

$$F(x, y, c) = 0.$$

By this equation, for each fixed value of c, a curve in the xy-plane is defined. If c varies, we have an infinite number of curves. The totality of these curves is called *family of curves*. In many engineering applications, there is also another family of curves that intersect each member of the given family at right angles. We say that these two families are *mutually orthogonal* and the curves of the second family are called the *orthogonal trajectories*.

For finding the orthogonal trajectories of a family of curves defined by (2.3), we realize the following procedure. First, we find a differential equation associated with (2.3), i.e., $y' = f(x, y)$. Let $m = f(x, y)$, then $m' = -1/f(x, y)$, where m and m' are the slopes of these two families, respectively. Then, we conclude that the differential equation of the orthogonal trajectories will be

$$y' = -1/f(x, y). \tag{2.30}$$

The solution of (2.30) is the family of curves which are orthogonal to the curves of the given problem (2.3).

Example 2.40. Find the orthogonal trajectories of

$$y = ce^x.$$

Solution. Solving for c, we get $c = ye^{-x}$. Differentiating the given equation w.r.t. x, we obtain $m = y' = y$. Then, $m' = y' = -1/y$. The solution of this equation is

$$y^2 = c - 2x.$$

□

Example 2.41 (Temperature field). Curves of constant temperature $T(x, y) = c$ in a temperature field are called *isotherms*. Their orthogonal trajectories are the curves along which the heat will flow (in regions that are free of sources or sinks of heat and are filled with homogeneous medium). If the isotherms are given by $y^2 + 2xy - x^2 = c$, what is the heat flow?

Solution. Differentiation of the given equation yields

$$y' = \frac{x - y}{x + y}.$$

Therefore, the curves of flow can be found by solving

$$y' = \frac{y + x}{y - x}.$$

This is a homogeneous equation. Let $y = xz$, then $y' = z + xz'$. Substituting y and y' into this equation, we obtain

$$\frac{1 - z}{z^2 - 2z - 1} dz = \frac{1}{x} dx.$$

This is a separable equation. Integrating both sides, we get

$$-\frac{1}{2} \ln(z^2 - 2z - 1) = \ln(cx).$$

That is, $z^2 - 2z - 1 = c/x^2$. To write this expression in terms of y and x, it is sufficient to replace $z = y/x$.   □

Remark 2.42. For finding the orthogonal trajectories of a given family, it is convenient to use polar coordinates. Consider a curve given in polar form. In calculus, it has been shown that the angle ψ_1 between the radius vector OP and the tangent line to the curve at $P(r, \theta)$ is given by

$$\tan(\psi_1) = \frac{rd\theta}{dr}.$$

If two curves are orthogonal, then $\psi_2 = \psi_1 + \dfrac{\pi}{2}$, i.e.,

$$\tan(\psi_2) = -\cot(\psi_1) = -\frac{1}{\tan(\psi_1)}.$$

Hence for the second curve, we have

$$\frac{rd\theta}{dr} = -\frac{1}{\tan(\psi_1)}.$$

Therefore, if we write

$$M(r, \theta)dr + N(r, \theta)d\theta = 0, \tag{2.31}$$

then

$$\frac{rd\theta}{dr} = -\frac{rM(r, \theta)}{N(r, \theta)}.$$

Hence, the orthogonal trajectories must be the solutions of the equation

$$\frac{rd\theta}{dr} = \frac{N(r, \theta)}{rM(r, \theta)}.$$

□

Example 2.43. Find the orthogonal trajectories of the family of curves satisfying the polar equation

$$r^2 = a \sin(2\theta).$$

Solution. For the given curves, we have

$$\frac{rd\theta}{dr} = \frac{\sin(2\theta)}{\cos(2\theta)}.$$

Hence,

$$\frac{rd\theta}{dr} = -\frac{\cos(2\theta)}{\sin(2\theta)}.$$

Separating the variables, we obtain

$$-\frac{\sin(2\theta)}{\cos(2\theta)}d\theta = \frac{dr}{r}.$$

Integrating this equation, we get

$$r^2 = b \cos(2\theta).$$

□

Example 2.44. Show that the orthogonal trajectories of the family of curves determined by the polar equation

$$r = a(1 - 2\sin(\theta))$$

satisfy

$$r^2 = b\cos(\theta(1 + \sin(\theta))).$$

Solution. The differentiation of the given equation yields

$$\frac{rd\theta}{dr} = -\frac{1 - 2\sin(\theta)}{2\cos(\theta)}.$$

Then

$$\frac{rd\theta}{dr} = \frac{2\cos(\theta)}{1 - 2\sin(\theta)}.$$

It is not difficult to show that

$$r^2 = b\cos(\theta(1 + \sin(\theta))).$$

□

2.5 Exercises

Exercise 2.1. Solve the following differential equations:

(1) $xy^2dx + (y + 1)e^x dy = 0.$

(2) $\tan(y)dx + (x - \sin^2(y))dy = 0.$

(3) $\left(\sin(y) - \dfrac{x}{y}\right)y' = 1.$

(4) $x^2 dx = y(ydx - xdy)e^{y/x}.$

(5) $xdx + \sin^2\left(\dfrac{y}{x}\right)(ydx - xdy) = 0, \quad y(1) = 0.$

(6) $(\sin(\theta) - \cos(\theta))dr + r(\sin(\theta) + \cos(\theta))d\theta = 0.$

(7) $y(e^x + y)dx + (e^x + 2xy)dy = 0.$

(8) $y' = xe^{y-x^2}, \quad y(0) = 0.$

(9) $2\tan(y)\sin(2x)dx + (\sin(y) - \cos(2x))dy = 0.$

(10) $e^{y-x^2}\left(x - 3 + \dfrac{3}{x} - \dfrac{1}{x^2}\right)dx = \dfrac{y}{x^2}dy, \quad y(1) = 1.$

(11) $y\left(2x^2 - xy + y^2\right)dx - x^2\left(2x - y\right)dy = 0.$

(12) $x\,(3xy - 4y^3 + 6)\,dx + (x^3 - 6x^2y^2 - 1)\,dy = 0.$

(13) $(\cos(2r) - 2\theta\sin(2r) - 2r\theta^3)\,dr + (\cos(2r) - 3r^2\theta^2)\,d\theta = 0.$

(14) $y\,(\ln(x) - \ln(y))\,dx + (y - x\ln(x) + x\ln(y))\,dy = 0,\quad y(1) = 1.$

(15) $y\tan^{-1}\left(\dfrac{x}{y}\right)dx + \left[y - x\tan^{-1}\left(\dfrac{x}{y}\right)\right]dy = 0,\quad y(1) = 1.$

(16) $x' - 2x\sin(y) = -x\sqrt{x}\sin(y).$

(17) $x^3 y\,dx + (3x^4 - y^3)\,dy = 0.$

(18) $y^3\sec^2(x)dx = (1 - 2y^2\tan(x))dy.$

(19) $(y')^2 - yy' = e^x.$

(20) $xy' - y = \sqrt{(y')^2 - 1}.$

(21) $2x\,(2y\,dx + dx + dy) + y^3\,(3y\,dx + 5x\,dy) = 0.$

(22) $y' + (\tan(x) + \cot(x))y = \displaystyle\sum_{n=0}^{\infty}(-1)^n\tan^{2n-1}(x).$

(23) $y' + x\sin(2y) = 2xe^{-x^2}\cos^2(y).$

(24) $xy' + y = 3x\sqrt{y}.$

(25) $2y^2 - \ln(y) + 3x(y - 2x)y' = 0.$

(26) $xy' + y = 2x^3 yy'\ln(y).$

(27) $xy' - 3y = x^5\sqrt[3]{y}.$

(28) $xy' - y\,(\ln(y) - \ln(x) + 1) = 0.$

(29) $y\,(2x^2 - xy + 1)\,dx + (x - y)\,dy = 0.$

(30) $x(x - \sec(y)\ln(x))y' + x\tan(y) - y\sec(y) = 0.$

Exercise 2.2. By introducing a suitable change of variables, solve the following equations:

(1) $xe^y y' - e^y = \dfrac{2}{x}.$

(2) $y' = 2\cos^2(x + y).$

(3) $xy' - (\ln(xy) - 1) = 0.$

(4) $xy' = y + x^{n+1}\sec\left(\dfrac{y}{x}\right).$

(5) $y' + \sec(y)\cos(x) = \tan(y)\cot(x).$

(6) $3y^2 y' = 2\,(y^3 - 4) - 5x\,(y^3 - 4)^3.$

(7) $(\cos(y) - 3x)dx + x^2 \tan(y)dy = 0.$

(8) $\cos(y)(x + \cos(y))dx + x^2 \sin(y)dy = 0.$

(9) $\left(xy\sqrt{x^2 - y^2} + x\right)y' = y - x^2\sqrt{x^2 - y^2}.$

(10) $2y' = (x - y)\cot(x) - [(x - y)\cot(x)]^{-1} + 2.$

Exercise 2.3. The following ODEs are of the type

$$(ax + by + c)\,dx = (\tilde{a}x + \tilde{b}y + \tilde{c})dy$$

and can be transformed into a homogeneous form by choosing $x = X - \alpha$ and $y = Y - \beta$, where (α, β) are the coordinates of the intersection point of the lines $ax + by + c = 0$ and $\tilde{a}x + \tilde{b}y + \tilde{c} = 0$. Solve the following equations by such a transformation. If two lines are parallel, then let $ax + by = z$.

(1) $(x - 3y + 4)\,dx - (2x + y - 2)\,dy = 0.$

(2) $(2x + 3y + 5)\,dx - (3x + 2y - 6)\,dy = 0.$

(3) $(5x - 3y)\,dx + (4x + 2y + 1)\,dy = 0.$

(4) $(x - 2y + 2)\,dx - (x - 2y - 3)\,dy = 0.$

(5) $(y - 2x + 5)\,dx + (4x - 2y - 3)\,dy = 0.$

Exercise 2.4. Find the family of solutions of each of the following Riccati equations:

(1) $y' = \sin(x) + y\cos(x) + y^2, \quad y_1 = -\cos(x).$

(2) $xy' = y + x^3\left(y^2 - x^2\right), \quad y_1 = x.$

(3) $xy' - 2y = y^2 - x^4, \quad y_1 = x^2.$

(4) $y' + \sin(x)y = y^2 + \cos(x), \quad y_1 = \sin(x).$

(5) $xy' = x^2\ln(x)^2 - 2x\ln(x) + 1 + 2x^2(1 - \ln(x))y + x^2y^2, \quad y_1 = \ln(x).$

Exercise 2.5. The following ODEs have an integrating factor of the form $F = x^\alpha y^\beta$. By finding α and β, solve the equations:

(1) $x\left(x + y^2\right)y' - y(3x + 2y^2) = 0.$

(2) $y(2 - 3xy)dx - xdy = 0.$

(3) $ydx + x\left(x^2y - 1\right)dy = 0.$

(4) $\left(2x^2 + 5xy^3\right)y' + 4xy + 3y^4 = 0.$

(5) $2y\,(ydx + 2xdy) + x^3\,(5ydx + 3xdy) = 0.$

Exercise 2.6. Consider a non-exact ODE $M(x, y)dx + N(x, y)dy = 0$ and suppose $F = e^{\int g(z)dx}$ is an integrating factor for this equation. Show that:

1. If $z = xy$, then

$$\frac{M_y - N_x}{yN - xM}$$

   is an integrating factor for the ODE. Use this information to solve

$$y^2(xy + 2x^2 - 1)dx + x^2(y^2 + 2xy - 2)dy = 0.$$

2. If $z = xy$ and the ODE can be written in the form

$$yM(x, y)dx + xN(x, y)dy = 0,$$

   then

$$\frac{1}{xy(M - N)}$$

   is an integrating factor for this equation. Use this fact and solve

$$y\left(x^2y^2 + xy + 1\right)dx - x\left(x^2y^2 - xy + 1\right)dy = 0.$$

3. If $z = x + y$, then

$$\frac{M_y - N_x}{N - M}$$

   is an integrating factor for the ODE. Use this information to solve

$$\left(5x^2 + 2xy + 3y^3\right)dx + 3\left(x^2 + xy^2 + 2y^3\right)dy = 0.$$

4. If $z = x^2 + y^2$, then

$$\frac{M_y - N_x}{2(xM - yN)}$$

   is an integrating factor for the ODE. Use this fact and solve

$$x(y - 1)dx - (x^2 + y)dy = 0.$$

5. If $z = x^2 - y^2$, then

$$\frac{M_y - N_x}{2(yN + xM)}$$

is an integrating factor for the ODE. Use this information to solve

$$(y + y^2 - x^2) \, dx + (x^2 - x - y^2) \, dy = 0.$$

Exercise 2.7. An *envelope* of a family of curves in the plane is a curve that is tangent to each member of the family at some point. For example, the circle $x^2 + y^2 = 1$ is an envelope for $ax + by = 1$. It can be proved that a family of curves $y = G(x,c)$ has the envelope $y = g(x)$ if and only if there is a function in terms of c such that

$$G_c\,(x, c\,(x)) = 0 \quad \text{and} \quad g\,(x) = G\,(x, c\,(x)).$$

This indicates that for finding the envelope of a family of curves, it is enough to eliminate c from $y = G(x,c)$ and $G_c\,(x,c) = 0$. In fact, the envelope of a family of curves is a singular solution of (2.2).

Find the envelopes of the following differential equations:

(1) $y = xy' - 2(y')^2$.

(2) $y = xy' + 5y' \ln(y')$.

(3) $yy' = \sqrt{k^2 - y^2}$.

(4) $(x^2 - 1)\,(y')^2 - 2yy' + y^2 - 1 = 0$.

(5) $(y)^2 + x^3 y' - 2x^2 y = 0$.

Exercise 2.8. Find the orthogonal trajectories of the following families of curves:

(1) $y^2 = x^2\,(1 - cx)$.   (2) $r = a(1 + \sin(\theta))$.

(3) $y = 3x - 1 + ce^{-3x}$.   (4) $r^2 = a \sin(2\theta)$.

(5) $x^4\,(4x^2 + 3y^2) = c$.   (6) $r = a(\sec(\theta) + \tan(\theta))$.

(7) $x^n + y^n = c^n$.   (8) $r = \dfrac{a}{1 + 2\cos(\theta)}$.

(9) $y = c\,(\sec(x) + \tan(x))$.   (10) $r = a \sec(\theta) \tan(\theta)$.

Exercise 2.9. An ODE which models the hormone secretion is given by

$$h'(t) = 1 - \cos(2\pi t) - h(t), \quad h(0) = 1,$$

where $h(t)$ is the amount of a certain hormone in the blood at any time t. Solve this equation and find $h(t)$. What happens to the level of hormone in the blood as $t \to \infty$?

Table 2.1 Data for the Gompertz growth model

t	0	1	2	3	4	5	6	7
$x(t)$	3.000	3.625	4.321	4.785	5.231	5.488	5.601	5.821

Exercise 2.10. Find the response of an RL-circuit to a unit square wave. Assume the current is initially zero.

Exercise 2.11. Let us assume that the half-life of a radioactive substance is 24 years. How long does it take until 80 % of the original amount of the substance are disintegrated?

Exercise 2.12. Experiments show that the lines of an electric force of two opposite charges of the same strength at $(-1, 0)$ and $(1, 0)$ are the circles through $(-1, 0)$ and $(1, 0)$. Show that these circles have the representation $x^2 + (y - c)^2 = 1 + c^2$. Show that the equipotential lines (orthogonal trajectories) are the circles $(x + c)^2 + y^2 = 1 + c^2$.

Exercise 2.13. In fluid flow, the path of a particle of fluid is called *streamline* and the orthogonal trajectories of these lines are called *equipotential lines*. If we assume that the streamlines are $xy = c$, find the equipotential lines.

Exercise 2.14. A bacteria culture is known to grow at a rate proportional to the present amount. After 1 h, 500 strands of the bacteria are observed, and after 3 h 2,500 strands. Find an expression for the approximate number of strands of the bacteria at any time t. Approximate the number of strands of the bacteria which were originally in the culture.

Exercise 2.15. Gompertz growth model for a population may be written as

$$x'(t) = -kx \ln\left(\frac{x}{N}\right),$$

where $x(t)$ is the population at any time t, and k and N are positive constants. This model has been employed quite successfully in the modeling of the growth of animal tumors. Table 2.1 shows the size of the tumor in dependence of the time.

Apply Gompertz growth model to this data, and obtain estimates for k and N. How long does it take for the tumor to grow 90 % of its maximum possible size?

Exercise 2.16. In actuarial science, a commonly used model for the growth of a population $P(t)$ is the general Gompertz equation given by

$$P'(t) = P(a - b \ln(P)),$$

where a and b are constants. Solve this equation such that $P(0) = P_0$. What is the long-term behavior of $P(t)$?

Exercise 2.17. A mass m is accelerated by a time-varying force $e^{-\alpha t}v^3$, where v is its velocity. It also experiences a resistive force βv, where β is a constant, owing to its motion through the air. The equation of motion of a mass is therefore

$$mv' = e^{-\alpha t}v^3 - \beta v.$$

Find an expression for the velocity v of the mass as a function of time, where the initial velocity v_0 is given.

Reference

1. Hartman, P.: Ordinary Differential Equations. Birkhäuser, Boston/Basel/Stuttgart (1982)

Chapter 3
Second-Order Differential Equations

3.1 Linear Differential Equations

Recall that the general form of an ODE is

$$f(x, y, y', y'', \ldots, y^{(n)}) = 0, \quad x \in [a, b]. \tag{3.1}$$

If we choose $n = 2$ in formula (3.1), then the equation

$$f(x, y, y', y'') = 0, \quad x \in [a, b], \tag{3.2}$$

is a second-order differential equation. For the first-order ODEs, we found that their solutions contain one arbitrary constant, but for the second-order ODEs, the solutions must contain two arbitrary constants because two integrations are required to obtain these solutions. In general, finding the family of solutions for (3.2) is difficult and in most cases impossible, in particular when the equation is nonlinear. For this reason and the fact that linear equations are very often used in engineering and in the applied sciences, in this chapter we will deal with linear ODEs. But before discussing the solution of linear ODEs, let us consider two special cases of nonlinear equations.

Case 1. Suppose (3.2) does not contain y. Hence, we have

$$f(x, y', y'') = 0. \tag{3.3}$$

The substitution $y' \equiv p$ can be used to reduce the order of the equation, i.e., to transform it into a first-order ODE. Since $y'' = p'$, we get

$$f(x, p, p') = 0.$$

Case 2. Suppose (3.2) does not contain x, i.e., it is an autonomous ODE. So we have

$$f(y, y', y'') = 0. \tag{3.4}$$

M. Hermann and M. Saravi, *A First Course in Ordinary Differential Equations: Analytical and Numerical Methods*, DOI 10.1007/978-81-322-1835-7_3, © Springer India 2014

Let $y' \equiv p(y)$, then $y'' = \dfrac{dp}{dy}\, p$. Therefore, (3.4) can be written as

$$f\left(y, p, \frac{dp}{dy}\, p\right) = 0.$$

This is an equation of first order and may be solved for p.

Example 3.1. Solve $e^{y'} y'' = 2x$.

Solution. The given equation does not contain y, i.e., the equation belongs to Case 1. We set $y' \equiv p$. Hence, $y'' = p'$ and we get $e^p p' = 2x$. Integrating both sides yields $e^p = x^2 + c_1$. Thus, we obtain $p = \ln\left(x^2 + c_1\right)$. Now using $p = y' = \ln\left(x^2 + c_1\right)$ and integrating by part yields

$$y(x) = x \ln\left(x^2 + c_1\right) - 2x + 2\sqrt{c_1}\, \tan\left(\frac{x}{\sqrt{c_1}}\right) + c_2.$$

$\square$

Example 3.2. Solve $3yy'y'' = (y')^3 - 1$.

Solution. The given equation does not contain x, i.e., the equation belongs to Case 2. Let us set $y' = p(y)$. Hence, $y'' = \dfrac{dp}{dy}\, p$. Inserting this into the equation, we get $3yp^2 p' = p^3 - 1$. Obviously, the variables are separable, and we may write

$$\frac{3p^2 dp}{p^3 - 1} = \frac{dy}{y}.$$

Integrating both sides, we obtain $p^3 = 1 + c_1 y$, i.e., $p = \sqrt[3]{1 + c_1 y}$. Now write

$$p = y' = \sqrt[3]{1 + c_1 y}.$$

Once more integration leads to

$$(1 + c_1 y)^{2/3} = \frac{2c_1}{3}(x + c_2).$$

$\square$

What makes the solution of these ODEs (in both cases) difficult is the integration. For example, let us change in Example 3.2 the equation to $3y'y'' = (y')^3 - 1$. Although in the first step of the solution we get $\ln\left((y')^3 - 1\right) = y + c_1$, the next integration is impossible. We have to use a numerical method (see Chap. 7) to obtain the solution. Let us consider another example.

Example 3.3 (See also [5]). Solve

$$y'' = -\lambda e^y, \quad y(0) = y(1) = 0.$$

Solution. This is the Bratu-Gelfand equation in one-dimensional planar coordinates, which arises from a simplification of the solid fuel ignition model in thermal combustion theory, physical applications ranging from chemical reaction theory, radiative heat transfer, and nanotechnology to the expansion of the universe.

Set $y' \equiv p$, then $y'' = \dfrac{dp}{dy} p$. Hence, the equation becomes

$$\frac{dp}{dy} p = -\lambda e^y, \quad \text{i.e.} \quad p \, dp = -\lambda e^y dy.$$

Integration leads to

$$\frac{p^2}{2} = -\lambda e^y + c_1.$$

This can be written as $p = \pm\sqrt{a^2 - 2\lambda e^y}$, where $a^2 \equiv 2c_1$.
Let $a^2 - 2\lambda e^y \equiv z^2$. Then $-2\lambda y' e^y = 2z'z$, i.e.,

$$y' = \frac{2z'z}{a^2 - z^2}.$$

The substitution of $y' = p$ into this expression leads to $\dfrac{2z'z}{a^2 - z^2} = \pm z$. Thus,

$$\frac{2dz}{a^2 - z^2} = \pm dx.$$

The integration of both sides gives

$$\frac{2}{a} \tanh^{-1} \frac{z}{a} = c_2 \pm x.$$

This can be written as

$$z = \operatorname{atanh}\left(b \pm \frac{a}{2}x\right),$$

where $b \equiv (a \, c_2)/2$. Since $a^2 - 2\lambda e^y = z^2$, the solution in terms of y can be given by

$$\lambda e^y = a^2 \left[1 - \tanh^2\left(b \pm \frac{a}{2}x\right)\right] = \frac{a^2}{\cosh^2\left(b \pm \frac{a}{2}x\right)}.$$

It is not difficult to show that

$$y(x) = 2 \ln \frac{a}{\sqrt{2\lambda} \cosh\left(b \pm \frac{a}{2}x\right)}.$$

If we apply the boundary conditions, we encounter problems. For example, the first boundary condition yields $a = \sqrt{2\lambda} \cosh(b)$. Now, if we apply the second boundary condition, we obtain $y(x) = 0$, with $\lambda = 0$. This is the trivial solution. The exact solution, for $\lambda > 0$, is given by

$$y(x) = -2 \log \left\{ \frac{\cosh[0.5(x-0.5)\theta]}{\cosh(0.25\theta)} \right\},$$

provided that θ is the solution of the nonlinear algebraic equation

$$\theta = \sqrt{2\lambda} \cosh(0.25\theta).$$

The problem has none, one, or two solutions when $\lambda > \lambda_c$, $\lambda = \lambda_c$, and $\lambda < \lambda_c$, respectively, where λ_c satisfies the equation $1 = 0.25\sqrt{2\lambda} \sinh(0.25\theta)$.   □

More information about the solution of such nonlinear BVPs can be obtained by numerical methods. The numerical solution of BVPs by shooting methods will be discussed in the last chapter. The rest of this chapter and the next chapter are devoted to methods of solving second-order linear differential equations. Unfortunately, there is no specific formula to obtain the general solution, as was the case for first-order linear ODEs. However, there is an extensive mathematical theory for second-order linear differential equations.

The general form of a second-order linear ODE is given by

$$P(x) y'' + Q(x) y' + R(x) y = S(x), \quad x \in [a, b], \tag{3.5}$$

where $P(x)$, $Q(x)$, $R(x)$, and $S(x)$ are given continuous functions on $a < x < b$. Let in this interval $P(x) \neq 0$, then we may write (3.5) in the form

$$y'' + p(x) y' + q(x) y = r(x), \quad x \in [a, b], \tag{3.6}$$

where $p(x) \equiv \dfrac{Q(x)}{P(x)}$, $q(x) \equiv \dfrac{R(x)}{P(x)}$, and $r(x) = \dfrac{S(x)}{P(x)}$.

If $r(x) = 0$, we have

$$y'' + p(x) y' + q(x) y = 0, \tag{3.7}$$

which is a *homogeneous* or *complementary* equation. We note here that the concept of homogeneity is not related to its use in the previous chapter. In order to solve (3.6), it is necessary to solve the corresponding homogeneous equation.

The existence and uniqueness theorem for second-order linear ODEs can be stated as follows.

Theorem 3.4 (Fundamental theorem). *If the functions $p(x)$, $q(x)$, and $r(x)$ are continuous on some open interval $a < x < b$, then (3.6) with the initial conditions $y(x_0) = y_0$ and $y'(x_0) = y_1$, where $x_0 \in (a, b)$, has a unique solution $y = \phi(x)$ which exists on the whole interval.*

Proof. See, e.g., the book of Coddington and Levinson [1] ∎

If we write (3.7) in the form $L[y] = 0$, then

$$L[y] = y'' + p(x)y' + q(x)y = 0. \tag{3.8}$$

It is not difficult to show that L is a linear differential operator that is defined on the linear space of twice continuously differentiable functions $y(x)$ on the interval (a, b). For linear ODEs, it can easily be shown that

(i) $\quad L[y_1 + y_2] = L[y_1] + L[y_2]$, (ii) $\quad L[c\,y] = c\,L[y]$,

where y_1 and y_2 are solutions of (3.7). These two properties are called the *linearity properties*.

Theorem 3.5. *If y_1 and y_2 are solutions of the differential equation (3.7), then their linear combination $y(x) = c_1 y_1(x) + c_2 y_2(x)$ is also a solution of this equation.*

Proof. To show the claim, we substitute

$$y = c_1 y_1(x) + c_2 y_2(x) \tag{3.9}$$

into the ODE (3.7). The result is

$$
\begin{aligned}
L[c_1 y_1 + c_2 y_2] &= [c_1 y_1 + c_2 y_2]'' + p[c_1 y_1 + c_2 y_2]' + q[c_1 y_1 + c_2 y_2] \\
&= c_1 y_1'' + c_2 y_2'' + c_1 p y_1' + c_2 p y_2' + c_1 q y_1 + c_2 q y_2 \\
&= c_1 [y_1'' + p y_1' + q y_1] + c_2 [y_2'' + p y_2' + q y_2] \\
&= c_1 L[y_1] + c_2 L[y_2].
\end{aligned}
$$

Since $L[y_1] = 0$ and $L[y_2] = 0$, we have $L[c_1 y_1 + c_2 y_2] = 0$. Therefore, (3.9) satisfies the differential equation (3.7) for all values c_1 and c_2. ∎

Example 3.6. The functions $y_1 = e^{2x}$ and $y_2 = x^2 + x + \dfrac{1}{2}$ are solutions of the ODE

$$xy'' - 2(x + 1)y' + 4y = 0.$$

Show that $y = c_1 y_1 + c_2 y_2$ is a solution, too.

Solution. We have

$$y' = c_1 y_1' + c_2 y_2' = 2c_1 e^{2x} + c_2 (2x + 1), \quad y'' = 4c_1 e^{2x} + 2c_2.$$

Substituting this into the ODE yields

$$x \left(4c_1 e^{2x} + 2c_2\right) - 2 (x + 1) \left[2c_1 e^{2x} + c_2 (2x + 1)\right]$$
$$+ 4 \left[c_1 e^{2x} + c_2 \left(x^2 + x + \frac{1}{2}\right)\right]$$
$$= 4c_1 x e^{2x} + 2c_2 x - 4c_1 x e^{2x} - 4c_1 e^{2x} - 4c_2 x^2 - 6c_2 x - 2c_2$$
$$+ 4c_1 e^{2x} + 4c_2 x^2 + 4c_2 x + 2c_2$$
$$= 0.$$

□

Now we want to consider the following question. Is it possible to find constants c_1 and c_2 such that the linear combination (3.9) fulfills the initial conditions

$$y(x_0) = y_0, \quad y'(x_0) = y_0', \tag{3.10}$$

too, where $a < x_0 < b$? If so, then c_1 and c_2 must satisfy the equations

$$c_1 y_1(x_0) + c_2 y_2(x_0) = y_0,$$
$$c_1 y_1'(x_0) + c_2 y_2'(x_0) = y_0'. \tag{3.11}$$

Using Cramer's rule, we obtain

$$c_1 = \frac{W_{c_1}}{W}, \quad c_2 = \frac{W_{c_2}}{W}, \tag{3.12}$$

where

$$W = \begin{vmatrix} y_1(x_0) & y_2(x_0) \\ y_1'(x_0) & y_2'(x_0) \end{vmatrix}, \quad W_{c_1} = \begin{vmatrix} y_0 & y_2(x_0) \\ y_0' & y_2'(x_0) \end{vmatrix}, \quad W_{c_2} = \begin{vmatrix} y_1(x_0) & y_0 \\ y_1'(x_0) & y_0' \end{vmatrix}.$$

With these values of c_1 and c_2, the expression (3.9) satisfies the ODE (3.7) as well as the initial conditions (3.10). To guarantee that c_1 and c_2 are well determined by (3.12), we have to assume that the denominator is not zero. For both constants, the denominator is the same, namely,

$$W = \begin{vmatrix} y_1(x_0) & y_2(x_0) \\ y_1'(x_0) & y_2'(x_0) \end{vmatrix} = y_1(x_0)y_2'(x_0) - y_1'(x_0)y_2(x_0). \tag{3.13}$$

The determinant W is called *Wronskian determinant* of the solutions $y_1(x)$ and $y_2(x)$. Sometimes we use the more detailed notation $W(y_1, y_2)(x_0)$ to indicate that the expression on the right-hand side of (3.13) is a function of y_1 and y_2 and is evaluated at $x = x_0$. Summarizing, we have the following theorem.

Theorem 3.7. *Assume that y_1 and y_2 are two solutions of the ODE (3.7). If and only if it holds*

$$W(y_1, y_2)(x_0) = y_1(x_0)y_2'(x_0) - y_1'(x_0)y_2(x_0) \neq 0,$$

then there exist constants c_1 and c_2 such that the linear combination (3.9) satisfies the ODE (3.7) and the initial conditions (3.10).

The next theorem gives rise to call the expression $c_1 y_1 + c_2 y_2$ the general solution of the ODE.

Theorem 3.8. *Let y_1 and y_2 be two solutions of the ODE (3.7). If there exist a value x_0 with $W(y_1, y_2)(x_0) \neq 0$, then the solution manifold*

$$y(x) = c_1 y_1(x) + c_2 y_2(x),$$

where c_1 and c_2 are arbitrary constants, contains all solutions of the ODE (3.7)

Proof. Let $y(x)$ be any solution of (3.7). To prove the claim, we have to show that $y(x)$ is part of the linear combination $c_1 y_1 + c_2 y_2$, i.e., for a special choice of c_1 and c_2, we obtain from this linear combination the solution $y(x)$.

Let x_0 be a value for which $W(y_1, y_2)(x_0) \neq 0$. We evaluate $y_0 \equiv y(x_0)$, $y_0' \equiv y'(x_0)$ and consider the initial value problem

$$y'' + p(x)y' + q(x)y = 0, \quad y(x_0) = y_0, \quad y'(x_0) = y_0'. \qquad (3.14)$$

Obviously, the function y solves the initial value problem. On the other hand, Theorem 3.7 and $W(y_1, y_2)(x_0) \neq 0$ imply that constants c_1 and c_2 can be found such that $c_1 y_1 + c_2 y_2$ is also a solution of the initial value problem (3.14). The exact values of c_1 and c_2 are determined by Eqs. (3.12). Since the uniqueness of the solution of the initial value problem (3.14) is guaranteed by the assumptions of Theorem 3.4, the two solutions must be the same. Thus, for the correct choice of c_1 and c_2, we have

$$y(x) = c_1 y_1(x) + c_2 y_2(x),$$

i.e., $y(x)$ is part of the solution manifold $c_1 y_1 + c_2 y_2$. Finally, our assumption that $y(x)$ is an arbitrary solution of (3.7) implies that each solution of this ODE must be part of this solution manifold. ∎

Theorem 3.8 states that under the assumption $W(y_1, y_2)(x_0) \neq 0$, the linear combination $c_1 y_1 + c_2 y_2$ contains all solutions of the ODE (3.7). Therefore, the expression

$$y = c_1 y_1 + c_2 y_2,$$

where c_1 and c_2 are arbitrary constants, is called the *general solution* of the ODE.
Solutions y_1 and y_2 for which the Wronskian determinant does not vanish constitute
a so-called *fundamental system* or *basis* for the ODE (3.7).

To find a fundamental system is often a difficult task, and the question arises
whether an ODE of the form (3.7) has always such a basis. An answer gives the
following theorem.

Theorem 3.9. *Consider the ODE (3.7) with continuous coefficients $p(x)$ and $q(x)$
on an open interval (a, b). Let x_0 be an arbitrary value in (a, b). Suppose y_1 is the
solution of (3.7) which satisfies the initial condition*

$$y_1(x_0) = 1, \quad y_1'(x_0) = 0,$$

and y_2 is the solution of (3.7) which satisfies the initial condition

$$y_2(x_0) = 0, \quad y_2'(x_0) = 1.$$

Then, y_1 and y_2 constitute a fundamental system of solutions for the ODE (3.7).

Proof. The existence of the functions y_1 and y_2 is guaranteed by Theorem 3.4.
To show that y_1 and y_2 constitute a fundamental system, we have to compute the
Wronskian determinant at x_0. We obtain

$$W(y_1, y_2)(x_0) = \begin{vmatrix} y_1(x_0) & y_2(x_0) \\ y_1'(x_0) & y_2'(x_0) \end{vmatrix} = \begin{vmatrix} 1 & 0 \\ 0 & 1 \end{vmatrix} = 1.$$

Thus, the Wronskian determinant does not vanish and y_1 and y_2 constitute a
fundamental system.   ◼

At the end of this section, we will consider the linear nth-order differential
equation

$$y^{(n)} + p_1(x)\, y^{(n-1)} + p_2(x)\, y^{(n-2)} + \cdots + p_{n-1}(x)\, y' + p_n(x)\, y = 0,$$

$$x \in [a, b]. \quad (3.15)$$

Note that the ODE (3.7) studied before is a special case of (3.15).

Let us consider n solutions $y_1(x), y_2(x), \ldots, y_n(x)$ of (3.15). Obviously, the
linear combination of these solutions

$$y(x) = c_1 y_1 + c_2 y_2 + \cdots + c_n y_n \quad (3.16)$$

is also a solution of this ODE. Usually, this property is referred to as the *principle
of superposition*.

The solutions $y_1(x), y_2(x), \ldots, y_n(x)$ are called *linearly independent* on an interval $[a, b]$ if there exist real numbers $c_1, c_2, \ldots c_n$ not all zero such that

$$c_1 y_1 + c_2 y_2 + \cdots + c_n y_n = 0.$$

A set of n linearly independent solutions $y_1(x), y_2(x), \ldots, y_n(x)$ of (3.15) on the interval (a, b) is called *basis* or *fundamental system* of the ODE (3.15) on this interval.

In the n-dimensional case, the generalization of the Wronskian determinant (3.13) is

$$W(y_1, y_2, \ldots, y_n)(x_0) = \begin{vmatrix} y_1 & y_2 & \cdots & y_n \\ y_1' & y_2' & \cdots & y_n' \\ \vdots & \vdots & & \vdots \\ y_1^{(n-1)} & y_2^{(n-1)} & \cdots & y_n^{(n-1)} \end{vmatrix}. \tag{3.17}$$

Theorem 3.8 can be transferred analogously to the n-dimensional case.

Finally, we have the following statement.

Theorem 3.10. *The nth-order linear homogeneous differential equation* (3.15) *possesses* n *linearly independent solutions. If* $y_1(x), y_2(x), \ldots, y_n(x)$ *are these solutions, then the general solution of* (3.15) *is given by* (3.16).

Proof. See, e.g., [2]. ☐

Example 3.11. Show that

$$y_1 = x^3 \text{ and } y_2 = x^2|x|, \quad x \in \mathbb{R},$$

are linearly independent but $W(y_1, y_2)(x_0) = 0$. Why?

Solution. We have

$$c_1 y_1 + c_2 y_2 = 0 \rightarrow c_1 x^3 + c_2 x^2 |x| = 0 \rightarrow c_1 = c_2 = 0.$$

Hence, y_1 and y_2 are linearly independent. But

$$W(y_1, y_2)(x_0) = \begin{vmatrix} y_1 & y_2 \\ y_1' & y_2' \end{vmatrix}_{x=x_0} = \begin{vmatrix} x^3 & x^2|x| \\ 3x^2 & 3x|x| \end{vmatrix}_{x=x_0} = 0 \text{ when } x_0 \neq 0.$$

For $x_0 = 0$, we have $W(y_1, y_2)(0) = \begin{vmatrix} 0 & 0 \\ 0 & 0 \end{vmatrix} = 0$, because

$$y_2'(0) = \lim_{h \to 0} \frac{h^2 |h| - 0}{h - 0} = 0.$$

We can now conclude that although y_1 and y_2 are linearly independent, they cannot be solutions of any linear second-order ODE.   □

Before introducing methods for solving inhomogeneous equations, the next section is devoted to the solution of homogeneous ODEs.

3.2   Solution Methods for Homogeneous Equations

As we mentioned in the previous section, the representation of the general solution of linear homogeneous ODEs in terms of elementary functions is difficult and often impossible. Fortunately, the general solution of linear ODEs with constant coefficients can easily be obtained, and we put in mind that the linear ODEs with constant coefficients are of great practical importance as well as of theoretical interest. But for linear ODEs with variable coefficients, a general form of the solution does not exist. In this section, we introduce methods that can be used for second-order linear ODEs with specific properties. Let us begin with the homogeneous case.

3.2.1   Method of Order Reduction

Recall the ODE (3.7)

$$y'' + p(x)y' + q(x)y = 0,$$

and suppose a solution, say $y_1(x)$, is known. Let $y_2(x) = u(x)\,y_1(x)$, then

$$y_2' = u'y_1 + u\,y_1', \quad y_2'' = u''y_1 + 2u'y_1' + u\,y_1''.$$

Substituting these expressions for y_2, y_2', and y_2'' into (3.7) and ordering the terms w.r.t. u, u', and u'' gives

$$u''y_1 + u'(2y_1' + py_1) + u(y_1'' + py_1' + qy_1) = 0.$$

Since y_1 is a solution of (3.7), we have $y_1'' + py_1' + qy_1 = 0$. Hence, we obtain

$$u''y_1 + u'(2y_1' + py_1) = 0.$$

One can separate the variables and obtain

$$\frac{u''}{u'} = -\left(\frac{2y_1'}{y_1 + p}\right).$$

Integrating twice leads to

$$u = \int \exp\left[-\int^x p(t)dt\right]/y_1^2 dx. \tag{3.18}$$

Therefore, the general solution will be

$$y(x) = y_1 \left\{c_1 + c_2 \int \exp\left[-\int^x p(t)dt\right]/y_1^2 dx\right\}. \tag{3.19}$$

Example 3.12.   Show that $y_1 = e^x$ is a solution of the ODE

$$(x + 1)y'' - (2x + 3)y' + (x + 2)y = 0.$$

Find the general solution.

Solution.   We have $y_1 = y_1' = y_1'' = e^x$. Substituting this into the ODE, we obtain

$$(x + 1)e^x - (2x + 3)e^x + (x + 2)e^x$$

$$= xe^x + e^x - 2xe^x - 3e^x + xe^x + 2e^x = 0.$$

Since $p(x) = -\dfrac{2x + 3}{x + 1}$, we can write

$$u = \int \exp\left(-\int^x p(t)dt\right)/y_1^2 dx = \int \frac{\exp\left(\int \dfrac{2x+3}{x+1}dx\right)}{\exp(2x)}dx$$

$$= \int \frac{\exp(2x + \ln(x + 1))}{\exp(2x)}dx = \int (x + 1)dx = \frac{(x + 1)^2}{2}.$$

Using (3.19), we come to

$$y(x) = e^x \left[c_1 + c_2(x + 1)^2\right].$$

$\square$

Example 3.13.   Find the general solution of equation

$$(1 - x\cot(x))y'' - xy' + y = 0,$$

if $y_1(x) = x$ is given.

Solution.   We have $p(x) = -\dfrac{x}{1 - x\cot(x)}$.

Hence,

$$u = \int \frac{\exp\left(\int^x \frac{t}{1 - t\cot(t)}\,dt\right)}{x^2}\,dx = \int \frac{\exp\left(\int^x \frac{\sin(t)}{\sin(t) - t\cos(t)}\,dt\right)}{x^2}\,dx$$

$$= \int \frac{\exp\left(\ln(\sin(x)) - x\cos(x)\right)}{x^2}\,dx = \int \frac{\sin(x)}{x^2}\,dx - \int \frac{\cos(x)}{x}\,dx.$$

Integrating by part for the first integral leads to

$$\int \frac{\sin(x)}{x^2}\,dx = -\frac{\sin(x)}{x} + \int \frac{\cos(x)}{x}\,dx.$$

Hence,

$$u = -\frac{\sin(x)}{x} + \int \frac{\cos(x)}{x}\,dx - \int \frac{\cos(x)}{x}\,dx = -\frac{\sin(x)}{x}.$$

We have $y_2 = uy_1 = -\sin(x)$. Thus,

$$y(x) = c_1 x + c_2 \sin(x).$$

□

Example 3.14. If $y_1 = x^2$ is given for

$$x^2 y'' + xy' - 4y = 0, \quad x > 0,$$

find the general solution of this equation.

Solution. Since $p(x) = \frac{1}{x}$, we get

$$u = \int \frac{\exp\left(-\int^x \frac{1}{t}\,dt\right)}{x^4}\,dx = \int \frac{1}{x^5}\,dx = -\frac{4}{x^4}.$$

Hence, $y_2 = uy_1 = -\frac{4}{x^2}$, i.e.,

$$y(x) = c_1 x^2 + c_2 x^{-2}.$$

□

Remark 3.15. The following properties may be used for solving an ODE with the method of order reduction:

1. If $p(x) + xq(x) = 0$, then $y_1(x) = x$.
2. If $2 + 2xp(x) + x^2q(x) = 0$, then $y_1(x) = x^2$.
3. If $1 + p(x) + q(x) = 0$, then $y_1(x) = e^x$.
4. In general, if $m^2 + p(x)m + q(x) = 0$, then $y_1(x) = e^{mx}$.
5. Given the ODE $p(x)y''(x) + axy'(x) + by(x) = 0$.
   If $a + b = 0$, then $y_1(x) = x$.

Thus, in Example 3.13 we have $a + b = 0$; hence, $y_1(x) = x$. In Example 3.12, we have

$$1 + p(x) + q(x) = 1 - \frac{2x + 3}{x + 1} + \frac{x + 2}{x + 1} = 0;$$

hence, $y_1(x) = e^x$. ☐

Example 3.16. Find the general solution of

$$(x - 1)y'' - xy' + y = 0.$$

Solution. Since $p(x) + xq(x) = -\dfrac{x}{x - 1} + \dfrac{x}{x - 1} = 0$, we get $y_1(x) = x$. We also have

$$1 + p(x) + q(x) = 1 - \frac{x}{x - 1} + \frac{1}{x + 1} = 0.$$

Thus, the second solution is given by $y_2(x) = e^x$. Now, the general solution can be written in the form $y(x) = c_1 x + c_2 e^x$. ☐

Example 3.17. Solve $xy'' - 2(x + 1)y' + 4y = 0$.

Solution. We have

$$0 = m^2 + p(x)m + q(x) = m^2 - 2\frac{x + 1}{x}m + \frac{4}{x} = (m - 2)\left(m - \frac{2}{x}\right).$$

Since m is a real number, we get $m = 2$. Hence, $y_1(x) = e^{2x}$ and another solution can be found by the method of order reduction. One may show that $y_2(x) = -\dfrac{1}{4}(2x^2 + 2x + 1)$ and hence $y(x) = c_1 e^{2x} + c_2(2x^2 + 2x + 1)$. ☐

Remark 3.18. The method of order reduction can be generalized as follows.
   By the method of order reduction, if $y_2 = uy_1$, then

$$u = \int^x \frac{y_p}{y_1^2} dx, \tag{3.20}$$

where $y_p \equiv e^{-\int^x p\,dx}$ and y_1 is known.

By differentiating (3.20) with respect to x, we obtain

$$u' = \frac{y_p}{y'^2},$$

whence $y_1 = \sqrt{\dfrac{y_p}{u'}}$.

One can write this relation as

$$y_1 = \sqrt{\frac{y_p}{u'u^{1-1}}}.$$

Now, let $y_2 = u'' y_1$. Then it can easily be proved that two independent solutions of (3.2), for two different values of n, can be found by

$$y = \sqrt{\frac{y_p}{u'u^{n-1}}}. \tag{3.21}$$

In particular, if $y_2 = x^n y_1$, then

$$y = \sqrt{\frac{y_p}{x^{n-1}}}. \tag{3.22}$$

$\square$

Example 3.19. Find the general solution of

$$y'' - 2\cot(x)y' + (2\cot^2(x) + 1)y = 0.$$

Solution. It is clear that $y_p(x) = \sin^2(x)$. Therefore,

$$y(x) = \sqrt{\frac{\sin^2(x)}{x^{n-1}}} = x^{\frac{1-n}{2}} \sin(x).$$

By substituting y, y', and y'' into the equation, it can be seen that $n = \pm 1$. Therefore, $y_1(x) = \sin(x)$ and $y_2(x) = x\sin(x)$. Hence, we have the general solution $y(x) = c_1 \sin(x) + c_2 x \sin(x)$. $\square$

Example 3.20. Solve $2(x^3 + x^2)y'' - (x - 3x^2)y' + y = 0$.

Solution. We have $y_p(x) = \dfrac{\sqrt{x}}{(x+1)^2}$. Thus,

$$y(x) = \frac{x^{\frac{1}{2}(\frac{3}{2}-n)}}{x+1}.$$

Now, by finding y, y', and y'' and substituting them into the given equation we obtain $n = \pm\dfrac{1}{2}$. From these two values of n, we obtain $y_1(x) = \dfrac{\sqrt{x}}{x+1}$ and $y_2(x) = \dfrac{x}{x+1}$. Thus, $y(x) = c_1\dfrac{\sqrt{x}}{x+1} + c_2\dfrac{x}{x+1}$. $\qquad\square$

Remark 3.21. The equation

$$ax^2y'' + bxy' + c = 0 \tag{3.23}$$

is known as Cauchy-Euler equation. By (3.22), solutions of this equation are given by

$$y(x) = \sqrt{\frac{x^{-\frac{b}{a}}}{x^{n-1}}}. \tag{3.24}$$

By substituting y, y', and y'' into the Cauchy-Euler equation, it can be proved that the values of n can be specified by

$$n = \pm\sqrt{\frac{a^2 + b^2 - 2ab - 4ac}{a^2}}. \tag{3.25}$$

This idea can be extended to any Cauchy-Euler equation of the form

$$a_k x^k y^{(k)} + a_{k-1}x^{k-1}y^{(k-1)} + \cdots + a_1xy' + a_0y = 0,$$

with solutions

$$y(x) = \sqrt{\frac{x^{-\frac{a_{k-1}}{a_k}}}{x^{n-k}}}. \tag{3.26}$$

$\qquad\square$

Let us reconsider Example 3.14. The corresponding ODE is

$$x^2y'' + xy' - 4y = 0, \quad x > 0.$$

This ODE is a Cauchy-Euler equation. If we use (3.25) with $a = 1$, $b = 1$, and $c = -4$, we get $n = \pm 4$. Since $y_p(x) = x^{-1}$, we obtain $y_1(x) = x^2$ and $y_2(x) = x^{-2}$.

Example 3.22. Find the general solution of

$$x^3y''' + \frac{5}{2}x^2y'' - \frac{1}{2}xy' + \frac{1}{2}y = 0, \quad x > 0.$$

Solution. We have

$$y(x) = \sqrt{\frac{x^{-\frac{5}{2}}}{x^{n-2}}}.$$

Substituting y, y', y'', and y'' into this equation, we obtain $n = \pm\frac{3}{2}$, and $n = -\frac{5}{2}$. Hence, $y_1(x) = x$, $y_2(x) = \frac{1}{x}$, $y_3(x) = \sqrt{x}$. Thus,

$$y(x) = c_1 x + c_2 \frac{1}{x} + c_3 \sqrt{x}.$$

It may be asked, "what happens if the values of n are complex?" It does not make any difference. In that case, we obtain the solutions in complex form.   □

Example 3.23. Find the general solution of

$$x^2 y'' + xy' + 4y = 0, \quad x > 0.$$

Solution. The general solution of this equation is

$$y(x) = c_1 \cos(2\ln(x)) + c_2 \sin(2\ln(x)).$$

If we use (3.26), we obtain

$$y(x) = \sqrt{\frac{x^{-1}}{x^{n-1}}} = x^{-\frac{n}{2}}.$$

Substituting y, y', and y'' into the given equation, we obtain $n = \pm 4i$. Therefore, the general solution is

$$y(x) = c_1 x^{2i} + c_2 x^{-2i}.$$

But, by definition, $z^c = \exp(c\ln(z))$ where c is a complex number. Since $c = \pm 2i$ and x is real, the general solution is

$$y(x) = b_1 \exp(2i \ln(x)) + b_2 \exp(-2i \ln(x)).$$

That is,

$$y(x) = b_1[\cos(2\ln(x)) + i \sin(2\ln(x))] + b_2[\cos(2\ln(x)) - i \sin(2\ln(x))]$$
$$= c_1 \cos(2\ln(x)) + c_2 \sin(2\ln(x)),$$

where $c_1 \equiv b_1 + b_2$ and $c_2 \equiv i(b_1 - b_2)$.   □

Results in generalizing the method of order reduction show the efficiency of this method for determining solutions in closed form when $y_2 = x^n y_1$. Moreover, we will later show how the Cauchy-Euler equation of any order can be solved by using a change of variables. But the idea of generalizing the method of order reduction is not restricted to the Cauchy-Euler equation. It can be used for any second-order linear ODE that has a solution of the form $y_2 = x^n y_1$.

3.2.2 Exact Equations

The concept of exactness in first-order ODEs can be extended to second-order linear ODEs. We call the equation

$$P(x)y'' + Q(x)y' + R(x)y = 0, \tag{3.27}$$

exact if

$$P''(x) - Q'(x) + R(x) = 0. \tag{3.28}$$

If an equation is exact, then it can be proved that (3.27) may be written in the form

$$[P(x)y']' + [f(x)y]' = 0, \tag{3.29}$$

where $f(x) \equiv Q(x) - P'(x)$.

Integrating equation (3.29) gives $P(x)y' + f(x)y = c_1$. That is,

$$P(x)y' + [Q(x) - P'(x)]y = c_1. \tag{3.30}$$

This is a first-order linear ODE and can be easily solved.

Example 3.24. Find the general solution of

$$x(x-2)y'' + 4(x-1)y' + 2y = 0, \quad x \neq 0, 2.$$

Solution. We have

$$P''(x) - Q'(x) + R(x) = 2 - 4 + 2 = 0.$$

Therefore, the given equation is exact. Equation (3.30) implies

$$x(x-2)y' + 2(x-1)y = c_1.$$

By integration, we obtain $x(x-2)y = c_1 x + c_2$. Hence,

$$y(x) = \frac{c_1 x + c_2}{x(x-2)}.$$

The idea of exactness can be extended to linear ODEs of any order, and it can be proved that a necessary and sufficient condition for the equation

$$p_0(x)y^{(n)} + p_1(x)y^{(n-1)} + p_2(x)y^{(n-2)} + \cdots + p_{n-1}(x)y' + p_n(x)y = 0$$

to be exact is

$$p_n(x) - p'_{n-1}(x) + p''_{n-2}(x) + \cdots + (-1)^n p_0^{(n)}(x) = 0. \qquad (3.31)$$

□

Example 3.25. Solve the ODE

$$xy''' + (x^2 + x + 3)y'' + 2(1 + 2x)y' + 2y = 0.$$

Solution. We have

$$p_0(x) = x, \quad p_1(x) = x^2 + x + 3, \quad p_2(x) = 2(1 + 2x) \text{ and } p_3(x) = 2,$$

hence,

$$p_3(x) - p'_2(x) + p''_1(x) - p'''_0(x) = 2 - 4 + 2 - 0 = 0.$$

Thus, the ODE is exact and we can write

$$(xy'')' + \left[(x^2 + x + 3)y'\right]' + [(1 + 2x)y]' = 0.$$

By integration, we get

$$xy'' + (x^2 + x + 3)y' + (1 + 2x)y = c_1.$$

This is an inhomogeneous ODE which we will consider later. But if we choose $c_1 = 0$, then the corresponding homogeneous equation

$$xy'' + (x^2 + x + 2)y' + (1 + 2x)y = 0$$

is also an exact equation. Did we reach accidentally at an exact equation? It was not a coincidence. When we reduce the order in any exact ODE, the result is itself an exact equation. Therefore, we can use this property and solve the inhomogeneous equation. First, we write the inhomogeneous equation as

$$(xy')' + \left[(x^2 + x + 1)y\right]' = c_1.$$

Integrating this equation leads to

$$xy' + (x^2 + x + 1)y = c_1 x + c_2.$$

This is a linear first-order ODE and can be solved by the techniques described before. ☐

Remark 3.26. We may wish to find an integrating factor as we found it for linear *first-order* ODEs. But for linear *second-order* ODEs, it is difficult to find such a factor. Namely, if we suppose $F(x)$ is an integrating factor, then the equation $FPy'' + FQy' + Fy = 0$ must be exact. That is,

$$(FP)'' - (FQ)' + F = 0.$$

This leads to

$$PF'' + (2P' - Q)F' + (P'' - Q' + R)F = 0,$$

which is also a linear second-order ODE. The difficulty to solve this equation is the same as to determine the solution of the original equation. Hence, the difficulty of finding the general solution of equation (3.27) still remains. ☐

3.2.3 Pseudo-exact Equations

The ODE (3.27) is called *pseudo-exact* (see [4]) if

$$P'' - Q' + R = 1. \tag{3.32}$$

Suppose Eq. (3.27) is not exact and let $R = 1$. In order to make it exact (see Remark 3.26), we use an integrating factor F such that Eq. (3.27) is transformed into an exact one. That is,

$$FPy'' + FQy' + Fy = 0. \tag{3.33}$$

Now, to find F we have to solve $(FP)'' - (FQ)' + F = 0$. That is,

$$PF'' + (2P' - Q)F' + (P'' - Q' + 1)F = 0. \tag{3.34}$$

It can be seen that (3.34) is a pseudo-exact ODE.

Although the transformation of a non-exact into an exact ODE is difficult, and in many cases impossible, by the following procedure Eq. (3.27) can be transferred into a pseudo-exact form.

Dividing (3.27) by R, we obtain

$$\frac{P}{R}y'' + \frac{Q}{R}y' + y = 0. \tag{3.35}$$

If we now choose $\frac{P}{R} \equiv p$ and $\frac{Q}{R} \equiv q$, Eq. (3.35) can be written as

$$py'' + qy' + y = 0. \tag{3.36}$$

Differentiating (3.36) leads to

$$py''' + (q + p')y'' + (q' + 1)y' = 0. \tag{3.37}$$

Now, if we set $y' \equiv z$, we obtain

$$pz'' + (q + p')z' + (q' + 1)z = 0, \tag{3.38}$$

which is a pseudo-exact ODE. Equation (3.38) may be solved if (3.27) has closed form solutions. For example, let us reconsider Example 3.13. Here we have shown that the general solution of the equation

$$(1 - x\cot(x))y'' - xy' + y = 0$$

is given by $y = c_1 x + c_2 \sin(x)$. Now we apply the abovementioned procedure. First, we differentiate the equation and thereafter we set $y' = z$. We obtain

$$(1 - x\cot(x))z'' - \cot(x)(1 - x\cot(x))z' = 0.$$

It is not difficult to show that $z(x) = c_1 + c_2\cos(x)$. Thus, we can see that $y'(x) = c_1 + c_2\cos(x)$. By a simple integration, we obtain the general solution $y(x) = c_1 x + c_2\sin(x) + c_3$. It is easy to show that $c_3 = 0$ unless we deal with the inhomogeneous case.

Example 3.27. Find the general solution of

$$x(x - 2)u'' - 2(x - 1)u' + 2u = 0, \quad 0 < x < 2.$$

Solution. If we follow the above procedure, we come to

$$\frac{1}{2}x(x - 2)u''' = 0.$$

Integrating successively three times, we obtain $u(x) = c_1 x^2 + c_2 x + c_3$. By a simple substitution, we come to $c_3 = -c_2$. Hence, the general solution is

$$u(x) = c_1 x^2 + c_2(x - 1).$$

$\square$

Example 3.28. Solve $(1 + x^2)y'' - 4xy' + 6y = 0$.

Solution. The pseudo-exact form of the given ODE is

$$\frac{1}{6}(1 + x^2)z'' - \frac{1}{3}xz' + \frac{1}{3}z = 0.$$

Obviously, $z_1 = x$ and by the method of reduction, we get $z_2 = x^2 - 1$. Thus,

$$y_1' = x, \text{ i.e. } y_1(x) = \frac{x^2}{2} + c_1,$$

and

$$y_2' = x^2 - 1, \text{ i.e. } y_2(x) = \frac{x^3}{3} - x + c_2.$$

Substituting these expressions into the given equation leads to $c_1 = 6$ and $c_2 = 0$. Therefore, the general solution is

$$y(x) = b_1 \left(\frac{x^2}{2} - \frac{1}{6}\right) + b_2 \left(\frac{x^3}{3} - x\right) = a_0(1 - 3x^2) + a_1 \left(x - \frac{x^3}{3}\right),$$

where $a_0 = -\frac{1}{6}b_1$ and $a_1 = -b_2$.   □

Example 3.29. Solve $y'' - 2xy' + 2\lambda y = 0$.

Solution. This equation is known as *Hermite equation* and will be reconsidered in Chap. 6 in connection with the power series method. Obviously, solutions for $\lambda = 0$ and $\lambda = 1$ are $y(x) = y_0(x) = 1$ and $y(x) = y_1(x) = 2x$, respectively.
   Let $\lambda = 2$, i.e., we consider now the equation

$$y'' - 2xy' + 4y = 0.$$

Writing the equation as

$$\frac{1}{4}y'' - \frac{1}{2}xy' + y = 0,$$

and differentiating it, we obtain

$$\frac{1}{4}z'' - \frac{1}{2}xz' + \frac{1}{2}z = 0,$$

where $z \equiv y'$. Obviously $z(x) = z_2(x) = x$. Hence,

$$y(x) = y_2(x) = \frac{x^2}{2} + c.$$

Substituting $y(x)$ into the given Hermite equation yields $y_2(x) = 2x^2 - 1$.

If we choose $\lambda = 3$, we get $y'' - 2xy' + 6y = 0$. In the same way, we obtain

$$\frac{1}{6}z'' - \frac{1}{3}xz' + \frac{2}{3}z = 0.$$

This equation has the solution

$$z(x) = z_3(x) = 2x^2 - 1.$$

Since $z_3(x) = y'_3(x)$, the solution of the Hermite equation with $\lambda = 3$ is

$$y(x) = y_3(x) = \frac{2}{3}x^3 - x.$$

If we continue this technique, we obtain the following results:

$$\lambda = 4: \qquad y_4(x) = \frac{4}{3}x^4 - 4x^2 + 1,$$

$$\lambda = 5: \qquad y_5(x) = \frac{2}{15}x^5 - \frac{4}{3}x^3 + x,$$

$$\vdots \qquad \vdots$$

Note that in the literature the functions $y_\lambda(x)$ are known as Hermite polynomials. They can be computed by the following recursion formula:

$$y_{\lambda+1}(x) = 2xy_\lambda(x) - 2\lambda y_{\lambda-1}(x), \qquad \lambda = 1, 2 \ldots$$

Hermite polynomials play an important role in the Hermite interpolation (see, e.g., [3]).   □

Results from Examples 3.13, 3.27, and 3.28 in this work show the efficiency of this solution technique. In spite of the fact that in these examples we did not assume that the given equation has a closed form solution, with this technique such solutions may be found. In Example 3.29 it was shown that the Hermite equation can also be solved by this technique.

3.2.4   Equations with Constant Coefficients

Consider the linear ODE

$$L[y] = ay'' + by' + cy = 0, \tag{3.39}$$

where a, b, and c are real numbers. Since y, y', and y'' differ only by constant multiplicative factors, we seek solutions with the ansatz $y = e^{mx}$.

When this ansatz is substituted into (3.39), we obtain

$$e^{mx}(am^2 + bm + c) = 0.$$

Thus, e^{mx} is a solution of (3.39) if and only if

$$L(m) \equiv am^2 + bm + c = 0. \tag{3.40}$$

The equation $L(m) = 0$ is called the *characteristic equation* and has two roots, say m_1 and m_2. Then $y_1(x) = e^{m_1 x}$ and $y_2(x) = e^{m_2 x}$ are solutions of (3.39). The Wronskian of these solutions is

$$W(y_1, y_2) = \begin{vmatrix} e^{m_1 x} & e^{m_2 x} \\ m_1 e^{m_1 x} & m_2 e^{m_2 x} \end{vmatrix} = (m_2 - m_1)e^{(m_2 - m_1)x},$$

which is zero if and only if $m_2 = m_1$. Hence, if $m_2 \neq m_1$, then the general solution is

$$y(x) = c_1 e^{m_1 x} + c_2 e^{m_2 x}.$$

Remark 3.30. The comparison of (3.39) with (3.40) shows that one can get equation (3.40) directly from Eq. (3.39) without substitution. ☐

Example 3.31. Find the general solution of $3y'' - 2y' - y = 0$.

Solution. We have $3m^2 - 2m - 1 = 0$, where $m_1 = 1$ and $m_2 = -3/2$. Hence,

$$y(x) = c_1 e^x + c_2 e^{-\frac{3}{2}x}.$$

☐

Remark 3.32. Since

$$e^{\pm mx} = \cosh(mx) \pm \sinh(mx), \quad \text{where } m = -\frac{b}{2a} \pm \frac{\sqrt{b^2 - 4ac}}{2a} \equiv M \pm N,$$

we may wish to write the general solution of (3.39) in the form

$$y(x) = e^{Mx}(A \cosh(Nx) + B \sinh(Nx)). \tag{3.41}$$

☐

Example 3.33. Find the general solution of $y'' - 12y' + 32y = 0$.

Solution. We have $m^2 - 12m + 32 = 0$, where $M = 6$ and $N = 2$. Therefore,

$$y(x) = e^{6x}(A \cosh(2x) + B \sinh(2x)).$$

☐

Remark 3.34. If $m_2 \neq m_1$, but are complex numbers, i.e., $m = M \pm iN$, then the general solution of equation (3.39) may be written in the form

$$y(x) = e^{Mx}\left(c_1 e^{iNx} + c_2 e^{-iNx}\right). \tag{3.42}$$

Using Euler's identities, $e^{\pm iNx} = \cos(Nx) \pm i\sin(Nx)$, we may write (3.42) in the form

$$y(x) = e^{Mx}(A\cos(Nx) + B\sin(Nx)). \tag{3.43}$$

□

Example 3.35. Find the general solution of $y'' - 2y' + 2y = 0$.

Solution. We have $m^2 - 2m + 2 = 0$, where $m = 1 \pm i$. Thus,

$$y(x) = e^x(A\cos(x) + B\sin(x)).$$

□

Example 3.36. The undamped oscillatory motion of mass-spring combination of mass m and spring constant k can be described by the differential equation

$$m\frac{d^2y}{dt^2} + ky = 0.$$

Find the unique solution that satisfies $y(0) = y_0$ and $y'(0) = v_0$.

Solution. Let $\lambda \equiv k/m$. Then the characteristic equation $m^2 + \lambda = 0$ has the roots $m = \pm i\sqrt{\lambda}$. Hence,

$$y(t) = A\cos(\sqrt{\lambda}t) + B\sin(\sqrt{\lambda}t).$$

We apply the initial conditions given by $y(0) = y_0$ and $y'(0) = v_o$ and obtain $A = y_0$ and $B = v_0/\sqrt{\lambda}$.

Thus, the unique solution is

$$y(t) = y_0\cos(\sqrt{\lambda}t) + \frac{v_0}{\sqrt{\lambda}}\sin(\sqrt{\lambda}t).$$

□

Remark 3.37. If $m_2 = m_1 = m$, then $y_1 = e^{mx}$ and we may obtain a second solution by the method of order reduction. It can be shown that $y_2 = xe^{mx}$. Hence,

$$y(x) = (c_1 + c_2 x)e^{mx}.$$

□

Example 3.38. Solve $9y'' - 12y' + 4y = 0$.

Solution. The characteristic equation $9m^2 - 12m + 4 = 0$ can be written in the form $(3m - 2)^2 = 0$, i.e., $m_{1,2} = \dfrac{2}{3}$. Hence, the solution is

$$y(x) = (c_1 + c_2 x)\, e^{\frac{2}{3}x}.$$

$\square$

Let us now consider the linear homogeneous ODE of nth order with constant coefficients

$$y^{(n)}(x) + a_1 y^{(n-1)}(x) + \cdots + a_{n-1} y'(x) + a_n y(x) = 0, \qquad (3.44)$$

where $a_1, a_2, \ldots, a_n$ are real constants. The corresponding characteristic equation is

$$L(m) \equiv m^n + a_1 m^{n-1} + \cdots + a_{n-1} m + a_n = 0. \qquad (3.45)$$

According to the fundamental theorem of algebra, a polynomial of degree n has exactly n roots, counting multiplicity. Obviously, the roots can be both real and complex. There are four cases which must be considered.

(a) All roots of $L(m) = 0$ are real and distinct.

Assume that Eq. (3.45) has n distinct roots $m_1, m_2, \ldots, m_n$. In this case, the general solution of (3.44) can be written in the following form:

$$y(x) = c_1 e^{m_1 x} + c_2 e^{m_2 x} + \cdots + c_n e^{m_n x}, \qquad (3.46)$$

where $c_1, c_2, \ldots, c_n$ are real constants.

(b) The roots of $L(m) = 0$ are real and multiple.

Let $L(m) = 0$ have r distinct roots $m_1, m_2, \ldots, m_r$, the multiplicity of which, respectively, is equal to $k_1, k_2, \ldots, k_r$, where $k_1 + k_2 + \cdots + k_r = n$. Then, the general solution of (3.44) has the form

$$y(x) = \left(c_1 + c_2 x + \cdots + c_{k_1} x^{k_1 - 1}\right) e^{m_1 x} + \cdots$$

$$+ \left(c_{n-k_r+1} + c_{n-k_r+2} x + \cdots + c_n x^{k_r - 1}\right) e^{m_r x}. \qquad (3.47)$$

It is seen that (3.47) has exactly k_i terms corresponding to each root m_i of multiplicity k_i. These terms are formed by multiplying x to a certain degree by $e^{m_i x}$. The degree of x varies in the range from 0 to $k_i - 1$.

(c) The roots of $L(m) = 0$ are complex and distinct.

The complex roots of $L(m) = 0$ appear in the form of conjugate pairs of complex numbers

$$m_{1,2} = \alpha \pm i\beta, \quad m_{3,4} = \gamma \pm i\delta, \ldots. \tag{3.48}$$

In this case the general solution of (3.44) reads

$$y(x) = e^{\alpha x}(c_1 \cos(\beta x) + c_2 \sin(\beta x))$$

$$+ e^{\gamma x}(c_3 \cos(\delta x) + c_4 \sin(\delta x))$$

$$+ \cdots \tag{3.49}$$

(d) The roots of $L(m) = 0$ are complex and multiple.

In this case each pair of complex conjugate roots $\alpha \pm i\beta$ of multiplicity k produces $2k$ particular solutions

$$e^{\alpha x}\cos(\beta x), \quad e^{\alpha x}\sin(\beta x), \quad e^{\alpha x}x\cos(\beta x), \quad e^{\alpha x}x\sin(\beta x), \ldots,$$

$$e^{\alpha x}x^{k-1}\cos(\beta x), \quad e^{\alpha x}x^{k-1}\sin(\beta x).$$

Now, the part of the general solution of (3.44) corresponding to a pair of complex conjugate roots is constructed as follows:

$$y(x) = e^{\alpha x}\left(c_1 \cos(\beta x) + c_2 \sin(\beta x)\right) + xe^{\alpha x}\left(c_3 \cos(\beta x) + c_4 \sin(\beta x)\right)$$

$$+ \cdots + x^{k-1}e^{\alpha x}\left(c_{2k-1} \cos(\beta x) + c_{2k} \sin(\beta x)\right). \tag{3.50}$$

In general, when $L(m) = 0$ has both real and complex roots of arbitrary multiplicity, the general solution of (3.44) is represented as the sum of the above solutions of the form (a)–(d).

Example 3.39. Find the general solution of $y''' + y'' - 5y' + 3y = 0$.

Solution. We have

$$m^3 + m^2 - 5m + 3 = 0 \quad \text{or} \quad (m-1)^2(m+3) = 0.$$

Thus, $m_1 = m_2 = 1, m_3 = -3$ and the general solution is

$$y(x) = (c_1 + c_2 x)e^x + c_3 e^{-3x}.$$

$\square$

Example 3.40. Solve the ODE

$$y^{(4)} - y^{(3)} + 2y' = 0.$$

Solution. The characteristic equation is

$$L(m) = m^4 - m^3 + 2m = 0.$$

The polynomial of the left-hand side can be written in the form

$$m(m^3 - m^2 + 2) = 0.$$

Thus, $m_1 = 0$ is a simple root. It can be easily seen that $m_2 = -1$ is another simple root. Dividing $m^3 - m^2 + 2$ by $m + 1$ yields $m^2 - 2m + 2$. The roots of the quadratic equation $m^2 - 2m + 2 = 0$ are $m_{3,4} = 1 \pm i$. Thus, the general solution of the given ODE is

$$y(x) = c_1 + c_2 e^{-x} + e^x(c_3 \cos(x) + c_4 \sin(x)).$$

$\square$

Remark 3.41. Recall that the Cauchy-Euler equation is given by

$$ax^2 y'' + bxy' + c = 0.$$

Let $x = e^t$, then $dx = e^t dt$. That is, $y' = \dfrac{1}{x}\dfrac{dy}{dt}$ and $y'' = \dfrac{1}{x^2}\left(\dfrac{d^2 y}{dt^2} - \dfrac{dy}{dt}\right)$.

Substitute y' and y'' into the Cauchy-Euler equation. We obtain

$$a\frac{d^2 y}{dt^2} + (b-a)\frac{dy}{dt} + cy = 0. \tag{3.51}$$

This is an equation with constant coefficients.

A more general form is $a(Ax + B)^2 y'' + b(Ax + B)y' + cy = 0$. This equation can be transformed into one having constant coefficients by letting $ax + b = e^t$. $\square$

Example 3.42. Find the general solution of

$$3(x - 2)^2 y'' - 5(x - 2)y' + 5y = 0.$$

Solution. Let $x - 2 = e^t$. Then, we obtain

$$3\frac{d^2 y}{dt^2} - 8\frac{dy}{dt} + 5y = 0.$$

The characteristic equation is $3m^2 - 8m + 5 = 0$; hence, we get $m_1 = 1$ and $m_2 = 5/3$. Thus, the general solution is

$$y(x) = c_1 e^t + c_2 e^{\frac{5}{3}t} = c_1(x - 2) + c_2(x - 2)^{\frac{5}{3}}.$$

$\square$

Example 3.43. The steady-state temperature T in a region bounded by two long coaxial cylinders satisfies

$$r \frac{d^2 T}{dr^2} + \frac{dT}{dr} = 0,$$

where r is the distance from the common axis. Verify that

$$T(r) = \frac{T_1 \ln \left(\frac{b}{r} \right) + T_2 \ln \left(\frac{r}{a} \right)}{\ln \left(\frac{b}{a} \right)}$$

is the solution of the corresponding boundary value problem which states that the cylinders are of radii a and b, $a < b$, and are kept at a constant temperature T_1 and T_2, respectively.

Solution. The given ODE is a Cauchy-Euler equation. It can be easily shown that

$$T(r) = c_1 + c_2 \ln(r).$$

Substituting $T(a) = T_1$ and $T(b) = T_2$ into the equation leads to

$$c_1 = \frac{T_1 \ln(b) - T_2 \ln(a)}{\ln \left(\frac{b}{a} \right)} \quad \text{and} \quad c_2 = \frac{T_2 - T_1}{\ln \left(\frac{b}{a} \right)}.$$

Now, a simple substitution implies the desired result.   □

3.2.5   Normal Form

Consider Eq. (3.7)

$$y'' + py' + qy = 0,$$

where p and q are continuous functions on some interval I. Suppose $y = uv$, then $y' = u'v + uv'$ and $y'' = u''v + 2u'v' + uv''$.

Substituting y, y', and y'' into the ODE, we obtain

$$vu'' + (2v' + pv)u' + (v'' + pv' + qv)u = 0. \tag{3.52}$$

Let $2v' + pv = 0$. Then

$$v = \exp \left[-\frac{1}{2} \int p(x) \, dx \right].$$

It is not difficult to show that

$$v'' + pv' + qv = q - \frac{1}{4}p^2 - \frac{1}{2}p'.$$

Hence, Eq. (3.52) reduces to

$$u'' + (q - \frac{1}{4}p^2 - \frac{1}{2}p')u. \tag{3.53}$$

Equation (3.53) is commonly known as the *normal form* of (3.7) and can easily be solved if

$$q - \frac{1}{4}p^2 - \frac{1}{2}p' = \text{const.}$$

Example 3.44. Solve $y'' - \frac{2}{x}y' + \left(1 + \frac{2}{x^2}\right)y = 0, \quad x \neq 0.$

Solution. We have

$$q - \frac{1}{4}p^2 - \frac{1}{2}p' = 1 + \frac{2}{x^2} - \frac{1}{x^2} - \frac{1}{x^2} = 1.$$

Thus, we have

$$u'' + u = 0, \quad \text{i.e. } u = A\cos(x) + B\sin(x).$$

We also have

$$v = \exp\left[-\frac{1}{2}\int p(x)dx\right] = \exp\left[\int \frac{1}{x}dx\right] = x.$$

Then the general solution is

$$y(x) = uv = x(A\cos(x) + B\sin(x)).$$

$\square$

Remark 3.45. It is clear that if

$$q - \frac{1}{4}p^2 - \frac{1}{2}p' = \frac{c}{(ax+b)^2}, \quad \text{where } a, b, c \in \mathbb{R},$$

then Eq. (3.53) is a Cauchy-Euler equation.

$\square$

3.3   Solution Methods for Inhomogeneous Equations

Consider the second-order ODE

$$y'' + p(x)y' + q(x)y = r(x), \tag{3.54}$$

where $p(x)$, $q(x)$, and $r(x)$ are continuous functions on $a < x < b$.

Let y_c be the solution of the homogeneous equation corresponding to (3.54). In the following, this solution is denoted as the *complementary solution*. Furthermore, let y_p be any particular solution of (3.54). Then it can be proved that $y = y_c + y_p$ is the solution of (3.54). We call this solution the general solution of (3.54).

Suppose we have determined the complementary solution. How can we find the particular solution? In this section we introduce a general method for finding the particular solution. However, for equations with constant coefficients, we discuss a much simpler method which is basic in engineering problems.

3.3.1   *Method of Undetermined Coefficients*

In the previous section, we learned how to solve the equation $L[y] = 0$ in the constant coefficient case. We now couple this with a method for finding a particular solution of $L[y] = r(x)$, where $r(x)$ is strictly restricted to the case

$$r(x) = e^{ax}(a_0 x^n + a_1 x^{n-1} + \cdots + a_n) \begin{cases} b_1 \cos(\beta x) \\ b_2 \sin(\beta x). \end{cases} \tag{3.55}$$

Consider

$$ay'' + by' + cy = r(x), \quad a, b, c \in \mathbb{R}. \tag{3.56}$$

Basically, we guess that the particular solution has the form

$$y_p(x) = e^{ax}(A_0 x^n + A_1 x^{n-1} + \cdots + A_n)(A \cos(\beta x) + B \sin(\beta x)). \tag{3.57}$$

Then we substitute y_p, y_p', and y_p'' into the given equation and compare the terms with those of $r(x)$. In this process different possibilities may occur. We discuss them with the help of some solved examples.

Example 3.46. Find the particular solution of $y'' - 3y' - 4y = 5e^{-4x}$.

Solution. A reasonable guess would be $y_p = Ae^{-4x}$. It holds

$$L[Ae^{-4x}] = 16Ae^{-4x} + 12Ae^{-4x} - 4Ae^{-4x} = 24Ae^{-4x}.$$

Now, if this guess is a solution, then we must have $24Ae^{-4x} = 5e^{-4x}$.
Hence, we obtain

$$A = \frac{5}{24}, \text{ i.e. } y_p = \frac{5}{24}e^{-4x}.$$

□

Now we reconsider this example with a small change.

Example 3.47. Find the particular solution of $y'' - 3y' - 4y = 5e^{4x}$.

Solution. Again we guess, $y_p = Ae^{4x}$. Since

$$L\left[Ae^{4x}\right] = 16Ae^{4x} - 12Ae^{4x} - 4Ae^{4x} = 0,$$

we must have $0 \cdot Ae^{4x} = 5e^{4x}$. Thus, it is impossible to find A. Where does it come to such a difficulty? The difficulty arises from the fact that e^{4x} is a solution of $L[y] = 0$. To avoid this difficulty, we guess $y_p = Axe^{4x}$. Why? We may claim that $r(x)$ and $y_1(x)$ are not linearly independent. Thus, to overcome this difficulty, we adopt the same procedure as we have done in the case $m_1 = m_2$ for solving $L[y] = 0$. Therefore, let $y_p(x) = Axe^{4x}$, then $y_p'(x) = A(4x+1)e^{4x}$ and $y_p''(x) = 8A(2x + 1)e^{4x}$. Substituting y_p, y_p', and y_p'' into the equation, we obtain

$$8A(2x + 1)e^{4x} - 3A(4x + 1)e^{4x} - 4Axe^{4x} = 5e^{4x}, \text{ i.e. } A = 1.$$

Thus, $y_p = xe^{4x}$. □

Remark 3.48. If the characteristic equation has multiple roots, then one may guess that $y_p = Ax^2e^{ax}$. □

Example 3.49. Find the general solution of $y'' - 4y' + 4y = 3e^{2x}$.

Solution. The corresponding characteristic equation $F(m) = 0$ has a double root $m_1 = m_2 = 2$; hence, we guess $y_p = Ax^2e^{2x}$. If we substitute y_p, y_p', and y_p'' into the equation and compare both sides, we obtain $A = 3$. Thus, $y(x) = (c_1 + c_2x)e^{2x} + 3x^2e^{2x}$. □

Example 3.50. Find the general solution of $y'' + 9y = 2\cos(3x)$.

Solution. Clearly, $y_1(x) = \cos(3x)$ and $y_2(x) = \sin(3x)$ are two independent solutions of the corresponding homogeneous equation. Since the functions $y_1(x) = \cos(3x)$ and $r(x) = 2\cos(3x)$ are linearly dependent, we guess that $y_p(x) = x(A\cos(3x) + B\sin(3x))$. Thus,

$$y_p'' = (6B - 9Ax)\cos(3x) - (6A + 9Bx)\sin(3x).$$

Substitute y_p and y_p'' into the equation and compare both sides. We come to $A = 0$ and $B = \frac{1}{3}$. Hence, $y_p = \frac{1}{3}x\sin(3x)$.

The corresponding general solution is

$$y(x) = y_c + y_p = c_1 \cos(3x) + c_2 \sin(3x) + \frac{1}{3}x \sin(3x).$$

□

Remark 3.51. In connection with the ansatz (3.57), we have the following fact: if $r(x)$ is a linear combination of terms, then y_p will also be a linear combination of the terms in $r(x)$.   □

Example 3.52. Find the particular solution of $y'' - 4y = 8x^2 - 4e^{-2x}$.

Solution. Since $r(x)$ contains a polynomial of degree 2 and an exponential function that is a solution of the corresponding homogeneous equation, we have

$$y_p(x) = Ax^2 + Bx + C + Dxe^{-2x}, \text{ i.e. } y_p''(x) = 2A - 4D(1 - x)e^{-2x}.$$

Substituting the above expressions into the given equation leads to $A = -2$, $B = 0$, $C = -1$, and $D = 1$. Then

$$y_p(x) = -2x^2 - 1 + xe^{-2x}.$$

□

Remark 3.53. If the inhomogeneous equation is given by

$$a_0 y^{(n)} + a_1 y^{(n-1)} + a_2 y^{(n-2)} + \cdots + a_{n-1} y' + a_n y = c,$$

where $a_0, a_1, \cdots, a_n \in \mathbb{R}$ and $c \in \mathbb{R}$, then $y_p(x) = \dfrac{c}{a_n}$. But if $a_n = 0$, then $y_p(x) = \dfrac{cx}{a_{n-1}}$. In general, if $a_n = a_{n-1} = \cdots = a_{n-k+1} = 0$, then $y_p(x) = \dfrac{cx^k}{k! a_{n-k}}$.   □

3.3.2 Method of Variation of Parameters

Consider Eq. (3.54). Let y_1 and y_2 be two linearly independent solutions of the homogeneous equation corresponding to (3.54). Then the general solution of this homogeneous equation is given by $y_c = c_1 y_1 + c_2 y_2$, where $c_1, c_2 \in \mathbb{R}$. This representation of the solution leads us to the idea to use the following ansatz for the particular solution of (3.54)

$$y_p = u_1 y_1 + u_2 y_2, \tag{3.58}$$

i.e., the parameters c_1 and c_2 are transformed into "varying" functions $u_1(x)$ and $u_2(x)$, respectively. Therefore, this technique is called *method of variation of parameters*. It is also known as *method of Lagrange*.

If so, then

$$y'_p = u_1 y'_1 + u_2 y'_2 + u'_1 y_1 + u'_2 y_2.$$

To determine uniquely the unknown functions $u_1(x)$ and $u_2(x)$, we need two conditions. The requirement that $y_p = u_1 y_1 + u_2 y_2$ satisfies the inhomogeneous ODE is the first condition. Thus, we still have a degree of freedom and we demand

$$u'_1 y_1 + u'_2 y_2 = 0. \tag{3.59}$$

Thus,

$$y'_p = u_1 y'_1 + u_2 y'_2 \quad \text{and} \quad y''_p = u'_1 y'_1 + u'_2 y'_2 + u''_1 y_1 + u''_2 y_2.$$

Substituting y_p, y'_p, and y''_p into (3.54) and simplifying the resulting equation, we obtain

$$u'_1 y'_1 + u'_2 y'_2 = r(x). \tag{3.60}$$

The system of equations (3.59) and (3.60) for the unknowns u'_1 and u'_2 may be solved by Cramer's rule. The corresponding determinants are

$$W(y_1, y_2) = \begin{vmatrix} y_1 & y_2 \\ y'_1 & y'_2 \end{vmatrix} = y_1 y'_2 - y_2 y'_1,$$

$$W_{y_1} = \begin{vmatrix} 0 & y_2 \\ r(x) & y'_2 \end{vmatrix} = -y_2 r(x), \quad W_{y_2} = \begin{vmatrix} y_1 & 0 \\ y'_1 & r(x) \end{vmatrix} = y_1 r(x).$$

Thus, the solution is

$$u'_1 = -\frac{y_2 r(x)}{W(y_1, y_2)}, \quad u'_2 = \frac{y_1 r(x)}{W(y_1, y_2)}.$$

Integrating these equations, we get

$$u_1 = -\int \frac{y_2 r(x)}{W(y_1, y_2)} dx + c_1, \quad u_2 = \int \frac{y_1 r(x)}{W(y_1, y_2)} dx + c_2. \tag{3.61}$$

Let us set $c_1 = c_2 = 0$ since we are looking for a particular solution. Substituting (3.61) into (3.58), we obtain the following particular solution

$$y_p = -y_1 \int \frac{y_2 r(x)}{W(y_1, y_2)} dx + y_2 \int \frac{y_1 r(x)}{W(y_1, y_2)} dx. \tag{3.62}$$

Example 3.54. Solve the differential equation $y'' - 2y' + y = e^x \tan(x) \sec^2(x)$.

Solution. It is not difficult to show that $y_c(x) = (c_1 + c_2 x)e^x$. Let $y_1(x) = e^x$ and $y_2(x) = xe^x$, then $W(y_1, y_2) = e^{2x}$.

We have

$$u_1(x) = -\int \frac{y_2 r(x)}{W(y_1, y_2)} dx = -\int \frac{x \tan(x) \sec^2(x) e^{2x}}{e^{2x}} dx$$

$$= -\int x \tan(x) \sec^2(x) dx = -\frac{1}{2} \left(x \tan^2(x) - \tan(x) + x \right),$$

$$u_2(x) = \int \frac{y_1 r(x)}{W(y_1, y_2)} dx = \int \frac{\tan(x) \sec^2(x) e^{2x}}{e^{2x}} dx$$

$$= \int \tan(x) \sec^2(x) dx = \frac{\tan^2(x)}{2}.$$

Therefore,

$$y_p(x) = u_1(x) y_1(x) + u_2(x) y_2(x)$$

$$= -\frac{e^x}{2} \left(x \tan^2(x) - \tan(x) + x \right) + \frac{xe^x \tan^2(x)}{2}$$

$$= \frac{e^x}{2} (\tan(x) - x).$$

$\square$

Example 3.55. Find the general solution of

$$(1 - x)y'' + xy' - y = 2(x - 1)^2 e^{-x}.$$

Solution. In a first step, we have to write the given ODE in the form (3.54). We obtain

$$y'' - \frac{x}{x-1} y' + \frac{1}{x-1} y = -2(x - 1)e^{-x}.$$

Clearly,

$$y_1(x) = x, \quad y_2(x) = e^x, \quad W(y_1, y_2) = (x - 1)e^x, \quad r(x) = -2(x - 1)e^{-x}.$$

We compute

$$u_1(x) = -\int \frac{y_2 r(x)}{W(y_1, y_2)} dx = \int \frac{e^x 2(x - 1)e^{-x}}{(x - 1)e^x} dx = \int 2e^{-x} dx = -2e^{-x},$$

$$u_2(x) = \int \frac{y_1 r(x)}{W(y_1, y_2)} dx = -\int \frac{x 2(x - 1)e^{-x}}{(x - 1)e^x} dx = -\int 2xe^{-2x} dx$$

$$= \left(x + \frac{1}{2} \right) e^{-2x}.$$

Thus, $y_p = u_1 y_1 + u_2 y_2 = \left(\dfrac{1}{2} - x\right) e^{-x}$. The general solution is

$$y(x) = y_c + y_p = c_1 x + c_2 e^x + \left(\frac{1}{2} - x\right) e^{-x}.$$

$\square$

Remark 3.56. In the method of variation of parameters, if we let $y_p = u\, y_1$, then we have the method of order reduction. That is, the method of order reduction can also be used for inhomogeneous equations. $\square$

Example 3.57. Find the general solution of $x y'' - y' + (1 - x) y = 1$.

Solution. We know (see Example 3.16) that $y_c(x) = c_1 x + c_2 e^x$. Choose $y_1(x) = e^x$. We assume $y(x) = u(x) y_1(x) = u(x) e^x$. Then

$$y' = (u' + u) e^x, \quad \text{i.e. } y'' = (u'' + 2u' + u) e^x.$$

Substituting y, y', and y'' into the given equation, we obtain

$$u'' + \left(1 - \frac{1}{x - 1}\right) u' = \frac{1}{x - 1}.$$

Suppose $u' = z$, then we have

$$z' + \left(1 - \frac{1}{x - 1}\right) z = \frac{1}{x - 1},$$

which is a linear first-order ODE. One can show that

$$z(x) = \frac{1}{x - 1} + c_1(x - 1) e^{-x}, \quad \text{i.e. } u'(x) = \frac{1}{x - 1} + c_1(x - 1) e^{-x}.$$

Integrating both sides leads to $u(x) = \ln(x - 1) - c_1 x e^{-x} + c_2$. Hence,

$$y(x) = u(x) y_1(x) = e^x \ln(x - 1) + c_1 x + c_2 e^x.$$

An interesting point in this solution is that $y_c = c_1 x + c_2 e^x$ appears in the general solution; hence, we may guess that $e^x \ln(x - 1)$ is a particular solution. $\square$

Let us consider another problem where formula (3.62) is useful. Given the ODE (3.54) subject to the following linear separated two-point boundary conditions

$$a_1 y(a) + a_2 y'(a) = 0, \quad b_1 y(b) + b_2 y'(b) = 0. \tag{3.63}$$

In (3.62) we set (b, x) as limits for the first integral and (a, x) as limits for the second integral. This yields

$$y = y_1 \int_b^x -\frac{y_2(\xi)r(\xi)}{W(y_1, y_2; \xi)} d\xi + y_2 \int_a^x \frac{y_1(\xi)r(\xi)}{W(y_1, y_2; \xi)} d\xi.$$

One may write this relation in the form

$$y = -\int_b^x \frac{y_1(x)y_2(\xi)r(\xi)}{W(y_1, y_2; \xi)} d\xi + \int_a^x \frac{y_2(x)y_1(\xi)r(\xi)}{W(y_1, y_2; \xi)} d\xi.$$

Let us change the sign in the first integral by changing the limit of integration (b, x) to (x, b). We obtain

$$y = \int_a^x \frac{y_2(x)y_1(\xi)r(\xi)}{W(y_1, y_2; \xi)} d\xi + \int_x^b \frac{y_1(x)y_2(\xi)r(\xi)}{W(y_1, y_2; \xi)} d\xi. \tag{3.64}$$

Now, we define

$$G(x, \xi) \equiv \begin{cases} \dfrac{y_2(x)y_1(\xi)}{W(y_1, y_2; \xi)}, & x \geq \xi, \\[3mm] \dfrac{y_1(x)y_2(\xi)}{W(y_1, y_2; \xi)}, & x < \xi. \end{cases} \tag{3.65}$$

Hence, (3.64) is equivalent to

$$y = \int_a^b G(x, \xi)r(\xi)d\xi. \tag{3.66}$$

The function $G(x, \xi)$ is called *Green's function*. It is clear that $G(x, \xi)$ must be continuous in $[a, b]$ and satisfies the boundary conditions (3.63).

Assume that y_1 and y_2 satisfy boundary conditions of the form

$$a_1 y_1(a) + a_2 y_1'(a) = 0, \quad b_1 y_2(b) + b_2 y_2'(b) = 0. \tag{3.67}$$

In (3.66) we use the second formula for $G(x, \xi)$ as $x \to a^+$ and the first formula as $x \to b^-$. This shows that G satisfies boundary conditions similar to (3.67) when it is regarded as a function of x. We notify that corresponding boundary conditions are also satisfied by y.

Example 3.58. Solve $y'' + y = 1$, $y(0) = y(\pi/2) = 0$.

Solution. Using the method of variation of constants, we obtain the following general solution of the homogeneous problem:

$$y_c = c_1 \sin(x) + c_2 \cos(x).$$

Thus, $y_1 = \sin(x)$ and $y_2 = \cos(x)$. The corresponding Wronskian is

$$W(y_1, y_2; \xi) = \begin{vmatrix} \sin(x) & \cos(x) \\ \cos(x) & -\sin(x) \end{vmatrix} = -(\sin^2(x) + \cos^2(x)) = -1.$$

Moreover, the right-hand side of the ODE is $r(x) \equiv 1$. Substituting y_1, y_2, r, and $W(y_1, y_2; \xi)$ into (3.64), we obtain

$$y = -\int_0^x \cos(x)\sin(\xi)d\xi - \int_x^{\pi/2} \sin(x)\cos(\xi)d\xi$$

$$= \cos(x)\big(\cos(x) - 1\big) - \sin(x)\big(1 - \sin(x)\big) = 1 - \cos(x) - \sin(x).$$

This is the unique solution of the given linear boundary value problem. □

In the case of higher-order linear inhomogeneous ODEs with constant coefficients

$$y^{(n)} + a_1 y^{(n-1)} + \cdots + a_{n-1} y' + a_n y = r(x), \tag{3.68}$$

where $a_1, a_2 \ldots, a_n$ are real numbers, and the right-hand side $r(x)$ is a continuous function on some interval $[a, b]$, the method of variation of parameters can be extended in the following way. As before, the superposition principle says that the general solution $y(x)$ of (3.68) is the sum of the general solution $y_c(x)$ of the corresponding homogeneous equation and a particular solution $y_p(x)$ of the inhomogeneous equation, i.e.,

$$y(x) = y_c(x) + y_p(x). \tag{3.69}$$

Assume that the general solution of the homogeneous ODE is known and given by

$$y_c(x) = c_1 y_1(x) + c_2 y_2(x) + \cdots + c_n y_n(x); \tag{3.70}$$

see Sect. 3.2.4. According to the method of variations of constants (Lagrange's method), for the particular solution of the inhomogeneous problem $y_p(x)$, the following ansatz is used:

$$y_p(x) = u_1(x)y_1(x) + u_2(x)y_2(x) + \cdots + u_n(x)y_n(x), \tag{3.71}$$

where $u_1, u_2, \ldots, u_n$ are functions which have to be determined. The derivatives of these functions are determined from the system of n equations

$$u_1'(x)y_1(x) + u_2'(x)y_2(x) + \cdots + u_n'(x)y_n(x) = 0$$
$$u_1'(x)y_1'(x) + u_2'(x)y_2'(x) + \cdots + u_n'(x)y_n'(x) = 0$$

$$\vdots \qquad\qquad (3.72)$$

$$u_1'(x)y_1^{(n-1)}(x) + u_2'(x)y_2^{(n-1)}(x) + \cdots + u_n'(x)y_n^{(n-1)}(x) = r(x).$$

The determinant of this system is the Wronskian of $y_1, y_2, \ldots, y_n$ forming a fundamental system of solutions. By the linear independence of these functions, the determinant is not zero and the system is uniquely solvable. The final expressions for the functions $u_1, u_2, \ldots, u_n$ can be found by integration.

Example 3.59. Find the general solution of the ODE

$$y''' + y' = \frac{1}{\cos(x)}$$

using the method of variation of parameters.

Solution. The characteristic equation of the homogeneous ODE is

$$F(m) \equiv m^3 + m = 0.$$

We write $m(m^2 + 1) = 0$ and obtain $m_1 = 0$ as a simple root. The other roots are $m_{2,3} = \pm i$. Consequently, the general solution of the homogeneous ODE is

$$y_c(x) = c_1 + c_2 \cos(x) + c_3 \sin(x),$$

where c_1, c_2, c_3 are arbitrary real numbers. According to the method of variation of parameters, we will consider the functions $u_1(x), u_2(x), u_3(x)$ instead of the numbers c_1, c_2, c_3 to construct a particular solution $y_p(x)$ of the inhomogeneous ODE in the form

$$y_p(x) = u_1(x)y_1(x) + u_2(x)y_2(x) + u_3(x)y_3(x). \qquad (3.73)$$

These functions satisfy equations (3.72), i.e.,

$$u_1'y_1 + u_2'y_2 + u_3'y_3 = 0$$
$$u_1'y_1' + u_2'y_2' + u_3'y_3' = 0 \qquad (3.74)$$
$$u_1'y_1'' + u_2'y_2'' + u_3'y_3'' = \frac{1}{\cos(x)},$$

where $y_1(x) = 1$, $y_2(x) = \cos(x)$, and $y_3(x) = \sin(x)$. Inserting these functions into (3.74) yields

$$u_1' + u_2' \cos(x) + u_3' \sin(x) = 0$$

$$- u_2' \sin(x) + u_3' \cos(x) = 0$$

$$- u_2' \cos(x) - u_3' \sin(x) = \frac{1}{\cos(x)}.$$

We use Cramer's rule (see, e.g., [3]) to determine the solution of this system. The main determinant is

$$W = \begin{vmatrix} 1 & \cos(x) & \sin(x) \\ 0 & -\sin(x) & \cos(x) \\ 0 & -\cos(x) & -\sin(x) \end{vmatrix} = \sin^2(x) + \cos^2(x) = 1.$$

The other three determinants are

$$W_{u_1'} = \begin{vmatrix} 0 & \cos(x) & \sin(x) \\ 0 & -\sin(x) & \cos(x) \\ \dfrac{1}{\cos(x)} & -\cos(x) & -\sin(x) \end{vmatrix} = \frac{1}{\cos(x)}(\cos^2(x) + \sin^2(x)) = \frac{1}{\cos(x)},$$

$$W_{u_2'} = \begin{vmatrix} 1 & 0 & \sin(x) \\ 0 & 0 & \cos(x) \\ 0 & \dfrac{1}{\cos(x)} & -\sin(x) \end{vmatrix} = -\frac{1}{\cos(x)}\cos(x) = -1,$$

$$W_{u_3'} = \begin{vmatrix} 1 & \cos(x) & 0 \\ 0 & -\sin(x) & 0 \\ 0 & -\cos(x) & \dfrac{1}{\cos(x)} \end{vmatrix} = -\sin(x)\frac{1}{\cos(x)} = -\tan(x).$$

Consequently, the derivatives u_1', u_2', and u_3' are given by

$$u_1' = \frac{W_{u_1'}}{W} = \frac{1}{\cos(x)}, \quad u_2' = \frac{W_{u_2'}}{W} = -1, \quad u_3' = \frac{W_{u_3'}}{W} = -\tan(x).$$

The integrals of these functions are

$$u_1(x) = \int \frac{dx}{\cos(x)} = \ln\left|\tan\left(\frac{x}{2} + \frac{\pi}{4}\right)\right| + c_4,$$

$$u_2(x) = -\int dx = -x + c_5,$$

$$u_3(x) = -\int \tan(x)dx = \ln|\cos(x)| + c_6.$$

Since we are only interested in a particular solution, we set $c_4 = c_5 = c_6 = 0$ and obtain

$$y_p(x) = \ln \left| \tan \left(\frac{x}{2} + \frac{\pi}{4} \right) \right| - x \cos(x) + \ln | \cos(x) | \sin(x).$$

Now, the general solution of the given ODE is

$$y(x) = y_c(x) + y_p(x)$$

$$= c_1 + c_2 \cos(x) + c_3 \sin(x)$$

$$+ \ln \left| \tan \left(\frac{x}{2} + \frac{\pi}{4} \right) \right| - x \cos(x) + \ln | \cos(x) | \sin(x).$$

$\square$

3.3.3   Method of Operators

A more sophisticated method to find a particular solution of a linear ODE with constant coefficients is the method of operators. The motivation behind this method is when the inverse operator is applied to $r(x)$, the particular solution results. Recall Eq. (3.56)

$$ay'' + by' + cy = r(x), \quad a, b, c \in \mathbb{R}.$$

This equation can be written in the form $F(D)y = r(x)$, where

$$F(D)y \equiv (aD^2 + bD + c)y = ay'' + by' + cy. \tag{3.75}$$

Note that D is an operator and must therefore always be followed by some expression on which it operates. Here, Dy means $Dy \equiv \dfrac{d}{dx}y$, but $uD \equiv u\dfrac{d}{dx}$. We define

$$D^2 y \equiv D \times Dy \equiv \frac{d}{dx} \left(\frac{dy}{dx} \right) = \frac{d^2 y}{dx^2} = y''.$$

Thus,

$$D^3 y = \frac{d^3 y}{dx^3} = y''', \quad \text{and more generally } D^n y = \frac{d^n y}{dx^n} = y^{(n)}.$$

The operator D satisfies the following algebraic laws:

- $D(y + z) = Dy + Dz.$
- $D^m(D^n y) = D^{(m+n)} y.$
- $D(zy) = zDy$ only when z is a constant.

We may write

$$y = \frac{1}{F(D)} r(x), \tag{3.76}$$

where $1/F(D)$ (inverse operator) is defined such that y in (3.76) is meaningful and that (3.75) is satisfied. That is, this method can be applied to linear ODEs of order n with constant coefficients. But before discussing some examples, we list some useful formulae based on the operator D:

1. It holds $De^{\lambda x} = \lambda e^{\lambda x}$ and $D^2 e^{\lambda x} = \lambda^2 e^{\lambda x}$. From which it can be seen that

$$F(D)e^{\lambda x} = (aD^2 + bD + c)e^{\lambda x} = (a\lambda^2 + b\lambda + c)e^{\lambda x}$$
$$= e^{\lambda x} F(\lambda). \tag{3.77}$$

2. It is not difficult to show (we leave the proof to self-study) that

$$(D - \lambda)^n (x^n e^{\lambda x}) = n! e^{\lambda x}. \tag{3.78}$$

3. We have

$$D^2(e^{\lambda x} y) = D(D(e^{\lambda x} y)) = D(\lambda e^{\lambda x} y + e^{\lambda x} Dy)$$
$$= \lambda^2 e^{\lambda x} y + \lambda e^{\lambda x} Dy + \lambda e^{\lambda x} Dy + e^{\lambda x} D^2 y$$
$$= e^{\lambda x} (D^2 + 2\lambda D + \lambda^2) y = e^{\lambda x} (D + \lambda)^2 y.$$

Similarly

$$D(e^{\lambda x} y) = \lambda e^{\lambda x} y + e^{\lambda x} Dy = e^{\lambda x} (D + \lambda) y.$$

Thus,

$$F(D)(e^{\lambda x} y) = aD^2(e^{\lambda x} y) + bD(e^{\lambda x} y) + ce^{\lambda x} y$$
$$= ae^{\lambda x} (D + \lambda)^2 y + be^{\lambda x} (D + \lambda) y + ce^{\lambda x} y$$
$$= e^{\lambda x} F(D + \lambda) y. \tag{3.79}$$

4. It holds

$$D^2 \cos(\alpha x + \beta) = D(D \cos(\alpha x + \beta)) = D(-\alpha \sin(\alpha x + \beta))$$
$$= -\alpha^2 \cos(\alpha x + \beta),$$
$$D^4 \cos(\alpha x + \beta) = D^2(D^2 \cos(\alpha x + \beta)) = D^2(-\alpha^2 \cos(\alpha x + \beta))$$
$$= (-\alpha^2)^2 \cos(\alpha x + \beta).$$

Thus,

$$
\begin{aligned}
F(D^2)\cos(\alpha x + \beta) &= (aD^4 + bD^2 + c)\cos(\alpha x + \beta) \\
&= \left(a(-\alpha^2)^2 + b(-\alpha^2) + c\right)\cos(\alpha x + \beta) \\
&= F(-\alpha^2)\cos(\alpha x + \beta).
\end{aligned}
\tag{3.80}
$$

5. Analogously, we obtain

$$
F(D^2)\sin(\alpha x + \beta) = F(-\alpha^2)\sin(\alpha x + \beta).
\tag{3.81}
$$

6. Finally, it can be shown that

$$
\frac{1}{D^2 + \gamma^2}\sin(\lambda x) = \frac{\sin(\lambda x)}{-\lambda^2 + \gamma^2}, \quad \gamma \neq \lambda,
$$

$$
\frac{1}{D^2 + \gamma^2}\cos(\lambda x) = \frac{\cos(\lambda x)}{-\lambda^2 + \gamma^2}, \quad \gamma \neq \lambda,
\tag{3.82}
$$

$$
\frac{1}{D^2 + \gamma^2}\sin(\lambda x) = -\frac{x\cos(\lambda x)}{2\gamma}, \quad \gamma = \lambda,
$$

$$
\frac{1}{D^2 + \gamma^2}\cos(\lambda x) = \frac{x\sin(\lambda x)}{2\gamma}, \quad \gamma = \lambda.
\tag{3.83}
$$

Under the assumption $F(\lambda) \neq 0$, we can conclude from 1 that

$$
\frac{1}{F(D)}e^{\lambda x} = \frac{e^{\lambda x}}{F(\lambda)}.
\tag{3.84}
$$

If $F(\lambda) = 0$, then $F(D)$ contains the factor $(D - \lambda)$. Let $F(D) = (D - \lambda)^n g(D)$, with $g(\lambda) \neq 0$. Thus, by (3.84) we have

$$
\frac{1}{F(D)}e^{\lambda x} = \frac{1}{(D - \lambda)^n g(D)}e^{\lambda x} = \frac{x^n e^{\lambda x}}{n!\,g(\lambda)}.
\tag{3.85}
$$

Example 3.60. Find the particular solution of $y'' + y = 5e^{3x}$.

Solution. Obviously, the operator form of the ODE is

$$
F(D)y \equiv (D^2 + 1)y = 5e^{3x}.
$$

Formally we can write

$$
y_p(x) = \frac{5e^{3x}}{F(D)}.
$$

Looking at the exponential function on the right-hand side of the ODE, we see that $\lambda = 3$. Then formula (3.84) says that

$$\frac{1}{F(D)}e^{3x} = \frac{e^{3x}}{F(3)}, \quad \text{where } F(3) = 10 \neq 0.$$

Thus, we get

$$y_p(x) = \frac{5e^{3x}}{F(3)} = \frac{e^{3x}}{2}.$$

□

Example 3.61. Find the particular solution of

$$D^3(D + 2)(D + 3)^4 y = 36e^{-3x}.$$

Solution. Looking at the right-hand side of the ODE, we see that $\lambda = -3$. The function $F(D)$ can be written as

$$F(D) = (D + 3)^4 g(D), \quad \text{where } g(D) = D^3(D + 2), \quad g(-3) \neq 0.$$

Thus, $n = 4$ and we can use formula (3.85) as follows:

$$y_p(x) = \frac{36e^{-3x}}{(D + 3)^4 g(D)} = \frac{36x^4 e^{-3x}}{27 \times 4!} = \frac{x^4 e^{-3x}}{18}.$$

□

Example 3.62. Find the particular solution of

$$(D^3 + 6D^2 + 11D + 6)y = 2\sin(3x).$$

Solution. We write

$$y_p(x) = \frac{1}{D^3 + 6D^2 + 11D + 6} \, 2\sin(3x).$$

Looking at the right-hand side of the ODE, we see that $\alpha = 3$ and $\beta = 0$. Using the relation (3.81), we conclude that $D^2 = -3^2$. Thus,

$$y_p(x) = \frac{1}{-(3^2) D + 6(-(3^2)) + 11D + 6} \, 2\sin(3x) = \frac{1}{D - 24} \sin(3x).$$

To obtain an expression with D^2 (this value is known), we apply the binomial formula and continue to write

$$y_p(x) = \frac{D + 24}{D^2 - 576} \sin(3x) = \frac{D + 24}{-(3^2) - 576} \sin(3x) = \frac{D + 24}{-585} \sin(3x).$$

Now, a simple differentiation yields

$$y_p(x) = -\frac{\cos(3x) + 8\sin(3x)}{195}.$$

☐

Example 3.63. Find the particular solution of $(D^2 + 4)y = x^2 - 6$.

Solution. We write

$$y_p(x) = \frac{1}{(D^2 + 4)}\left(x^2 - 6\right) = \left(D^2 + 4\right)^{-1}\left(x^2 - 6\right).$$

In order to be able to apply the power series of $(1 + x)^{-1}$, we transform the right-hand side as follows:

$$y_p(x) = \frac{1}{4}\left(1 + \frac{D^2}{4}\right)^{-1}\left(x^2 - 6\right)$$

$$= \frac{1}{4}\left(1 - \frac{D^2}{4} + \frac{D^4}{16} - \cdots\right)\left(x^2 - 6\right)$$

$$= \frac{1}{4}\left(-6 - \frac{1}{2} + x^2 + 0 + 0 + \cdots\right)$$

$$= \frac{1}{4}\left(-\frac{13}{2} + x^2\right).$$

☐

3.4   Exercises

Exercise 3.1. For the following ODEs, find the singular solution (if it exists) and the family of solutions:

(1) $yy'' + (y')^3 = (y')^2$.   (2) $x^4 y'' = y'(y' + x^3)$.

(3) $y'' = 1 + (y')^2$.   (4) $y'' = y'\cosec(x)$.

(5) $y'' = y'\cot(x)$.   (6) $x^2 y'' = (y')^2$.

(7) $2yy'' = 1 + (y')^2$.   (8) $x^2 y'' = y'(4x - 3y')$.

(9) $yy'' + (y+1)(y')^2 = 0$.   (10) $yy'' = (y')^2(1 - y'\sin(y) - yy'\cos(y))$.

Exercise 3.2. Solve the following ODEs:

(1) $y'' - 2y' + y = (1 - e^{-x})^{-2}.$ (2) $y'' + y = \operatorname{cosec}^2(x)\cot(x).$

(3) $y'' - 3y' + 2y = \cos(e^{-x}).$ (4) $y'' - 4y = \sin(e^{-2x}).$

(5) $y''' + y' = \tan(x).$ (6) $y^{(4)} - 16y = 0.$

(7) $y'' + 2y' + 2y = \dfrac{e^{-x}}{\sin(x)}.$ (8) $y'' - y = 2e^{2x} - e^{-x}.$

(9) $y'' + 2y' + y = 3e^x - 2e^{-x}.$ (10) $y'' + 9y = \sin(3x).$

(11) $y''' + 4y'' - 5y = 2x - 1.$ (12) $y^{(4)} + y = 2\cos(x) - \sin(2x).$

(13) $(D^5 - 4D^3)y = 24.$ (14) $(D^3 - 1)y = 4 - 3x^2.$

(15) $(D-2)^2(D+1)y = xe^{2x}.$ (16) $(D^2+2D+2)y = e^x\cos(x)+x^2.$

(17) $(D^2 - 1)y = x\sin(x) + (1 + x^2)e^x.$

(18) $D(D - 1)^2(D + 2)y = 2x - 5e^x.$

(19)   · $(D - 1)^2(D^2 + 1)y = 3e^x - 2\cos(x).$

(20) $(D^3 - 2D^2 + 4D - 3)y = -24e^x.$

Exercise 3.3. Find the general solution of following differential equations:

(1) $(1 + x^2)y'' + xy' - y = 0.$

(2) $x^3y'' - x^2y' - 3xy = 16\ln(x).$

(3) $(1 - x^2)y'' - 2xy' + 2y = 0.$

(4) $(1 + x^2)y'' + 2xy' = \dfrac{3}{x^2}.$

(5) $(x + 1)y'' + (3x + 4)y' + 3y = 3x.$

(6) $(x - 1)^2y'' - 2y = \ln\left(\sqrt{x - 1}\right).$

(7) $x^2y'' + xy' = \cos(\ln(x)) + x\sin(\ln(x)).$

(8) $x^2y'' - 2xy' + 2y = (\ln(x))^2 + (\ln(x))^2.$

(9) $y'' - 2\tan(x)y' - 10y = e^{-3x}\tan(x).$

(10) $x^2y'' + xy' - y = \ln(x).$

Exercise 3.4. Sometimes it is useful to change the variables so that the given ODE is transferred into one that can be solved. By using the change of variables introduced in this section, solve the following differential equations:

(1) $x^3y'' + 2x^2 = (xy' - y)^2,$ $y = -x\ln(t).$

(2) $(1 + x^2)y'' + xy' - 4y = 0,$ $y = \sinh(t).$

(3) $(1 - x^2)y'' - xy' + y = x,$ $x = \cos(t).$

(4) $y'' + (9e^{-2x} - 4)y = 0,$ $\qquad\qquad$ $u = 3e^{-x}.$

(5) $y'' + 4x^2 y = 0,$ $\qquad\qquad\qquad$ $u = x^2, \; y = u\sqrt{x}.$

(6) $y'' + 16xy = 0,$ $\qquad\qquad\qquad$ $u = \dfrac{8}{3}x\sqrt{x}, \; y = u\sqrt{x}.$

(7) $y'' - 4xy' + 4x^2 y = xe^{x^2},$ $\qquad$ $y = ue^{x^2}.$

(8) $x^4 y'' + 2x^3 y' + y = 2 + \dfrac{1}{x^2},$ $\qquad$ $x = -\dfrac{1}{u}.$

(9) $(1 - x^2)y'' - xy' + y = x,$ $\qquad$ $x = \sin(u).$

(10) $y'' - 2\tan(x)y' + 3y = 2\sec(x),$ $\quad$ $y = u\sin(x).$

Exercise 3.5. Let

$$y'' + py' + qy = (D^2 + pD + q)y = (D + u)(D + v)y = 0,$$

where u and v are functions of x. Since $(D + v)y = y' + vy$, we obtain $(D + u)(y' + vy) = 0$. That is, $y'' + (u + v)y' + (v' + uv)y = 0$. It is not difficult to show that $u + v = p$ and $v' + uv = q$. By eliminating u from this system of equations, we obtain $v' = q - pv + v^2$, which is a Riccati equation and may be solved. By this operator method, find the general solution of

$$y'' + 2xy' + (1 + x^2)y = 0.$$

Exercise 3.6. Consider the ODE

$$x^2 y'' + xy' + \left(x^2 - \frac{1}{4}\right)y = 0,$$

known as Bessel equation with index $\lambda = 1/2$.

1. Substitute $y = uz$ and choose z such that the coefficient of u' is zero.
2. Show with this technique that the general solution of this equation is

$$y(x) = \frac{1}{\sqrt{x}} (c_1 \cos(x) + c_2 \sin(x)).$$

Exercise 3.7. Find the particular solution of the following equations:

(1) $y^{(4)} - y' = 2x,$ $\qquad\qquad$ $0 < x < 1.$

(2) $y''' - y' = 3e^{2x},$ $\qquad\qquad$ $0 < x < 1.$

(3) $y''' - y' = \sec(x),$ $\qquad\qquad$ $-\dfrac{\pi}{2} < x < \dfrac{\pi}{2}.$

(4) $y^{(4)} + 2y'' + y = \sin(x),$ $\quad$ $0 < x < \dfrac{\pi}{2}.$

(5) $y''' - 4y'' + 3y' = x,$ $\qquad\quad$ $-1 < x < 1.$

Exercise 3.8.

1. Suppose $y'' + py' + qy = 0$ is a pseudo-differential equation. Prove if $y = ue^{-\int p\,dx}$ is a solution, then u satisfies $u'' - pu' + u = 0$.
2. Find a closed form solution for

$$y'' + xy' + 2y = 0.$$

Exercise 3.9. Rayleigh's equation is the nonlinear ODE given by

$$\frac{d^2x}{dt^2} + \left[\frac{1}{3}\left(\frac{dx}{dt}\right)^2 - 1 \right]\frac{dx}{dt} + x = 0.$$

Among other applications, it arises in the study of a violin string. Solve this equation for the following initial conditions:

(1) $x(0) = 1$, $\quad x'(0) = 0$.
(2) $x(0) = 0.1$, $\quad x'(0) = 0$.

Exercise 3.10. A classical problem of mechanical engineering is the hanging cable problem. The shape of a hanging cable of uniform density is governed by the ODE

$$\frac{d^2y}{dx^2} = \frac{d}{T}\sqrt{1 + \left(\frac{dy}{dx}\right)^2},$$

where d is the constant density and T is the tension in the cable at its lowest point. Solve this differential equation.

Exercise 3.11. The ODE

$$\frac{d^2y}{dx^2} = \left[\left(\frac{dy}{dx}\right)^2 + 1 \right]\frac{dx}{dt}$$

occurs in the analysis of satellite orbits. Find all solution of this equation.

Exercise 3.12. A particle with mass m, free to move along the x-axis, is attracted toward the origin with a force proportional to its distance from the origin. Find the motion if it starts from rest at $x = x_0$.

Exercise 3.13. The ODE $EIu^{(4)} = f(x)$ arises in the analysis of the static deflection of a *rigid* thin homogeneous beam of uniform cross section, where E is the modulus of elasticity of the beam material and I is the moment of inertia of the beam cross section about the so-called neutral axis. Find the deflection $u(x)$ of the transverse load (find only a particular solution).

Exercise 3.14. Any ODE of the form

$$\left(y^{(n)}\right)^{n_1} + a_1 \left(y^{(n-1)}\right)^{n_2} + \cdots + a_{n-2}\left(y''\right)^{n_{n-1}} + a_{n-1}\left(P(y)\right)^{n_n} = a_n,$$

where $a_1, a_2, \ldots, a_n$ are constants and $P(y)$ is a polynomial, can be reduced to the order $(n - 1)$. This can be achieved if we multiply this equation by y' and then integrate both sides. Solve the following initial value problems:

(1) $y'' = \lambda(y + y^2)$, $y(0) = y'(0) = 0$.
(2) $y'' = y + \lambda y^3$, $y(0) = y'(0) = 0$.
(3) $y'' = -\lambda e^y$, $y(0) = y'(0) = 0$.

Exercise 3.15. Solve the following BVPs by use of Green's function

(1) $y'' = x - 1$, $y'(0) = y(1) = 0$.

(2) $y'' = \cos(x)$, $y(0) + y'(1) = 0$.

(3) $y'' + 4y = -1$, $y(0) = 0$, $y(\pi/4) = 0$.

(4) $y'' = x^2$, $y(0) = 0$, $y(1) + y'(1) = 0$.

(5) $y'' + y = 1$, $y(0) = 1$, $y(\pi/4) + y'(\pi/4) = 2$.

References

1. Coddington, E.A., Levinson, N.: Theory of Ordinary Differential Equations. MacGraw-Hill, New York (1984)
2. Hartman, P.: Ordinary Differential Equations. Birkhäuser Verlag, Boston/Basel/Stuttgart (1982)
3. Hermann, M.: Numerische Mathematik, 3rd edn. Oldenbourg Verlag, München (2011)
4. Saravi, M.: A procedure for solving some second-order differential equations. Appl. Math. Lett. **25**(3), 408–411 (2012)
5. Saravi, M., Hermann, M., Kaiser, D.: Solution of Bratu's equation by He's variational iteration method. Am. J. Comput. Appl. Math. **3**(1), 4–48 (2013)

Chapter 4
Laplace Transforms

4.1 Introduction

One of the very useful tools in solving linear ODEs and the corresponding initial and boundary values problems is the method of Laplace transforms. Laplace transforms are well adapted for the treatment of inhomogeneous equations that result when systems involving discontinuous or periodic input functions are modeled. They also can be used to solve certain types of improper integrals and integral equations.

We are familiar with the operator D defined in Sect. 3.3.3, and we observed that using this operator many functions can be transformed into other functions. In this chapter, we introduce another class of transformations that can be defined by

$$T\{f(x)\} = \int_{-\infty}^{\infty} k(s,t) f(t)\, dt = F(s), \tag{4.1}$$

where the function $k(s,t)$ is called the *kernel* of the transformation. Here $T\{f\}$ represents an entire class of transformations, since if k is changed, a new transform is defined. If we let

$$k(s,t) = \begin{cases} 0, & t < 0 \\ e^{-st}, & t \geq 0. \end{cases} \tag{4.2}$$

then we have a member of this class that is called the *Laplace transform*. This chapter is devoted to this transform, but before dealing with it, we give two definitions which prepare the formulation of a sufficient condition for the existence of a Laplace transform.

Definition 4.1. A function $f(x)$ is said to be a *piecewise continuous function* on a closed interval $[a,b] \in \mathbb{R}$ if there exist a finite number of points $a = x_0 < x_1 < x_2 < \cdots < x_N = b$ such that $f(x)$ is continuous in each of the intervals (x_{i-1}, x_i) for $1 \leq i \leq N$ and has finite limits as x approaches the endpoints. $\quad\square$

M. Hermann and M. Saravi, *A First Course in Ordinary Differential Equations: Analytical and Numerical Methods*, DOI 10.1007/978-81-322-1835-7_4, © Springer India 2014

An example of a piecewise continuous function is the greatest integer function $f(x) = [x]$.

Definition 4.2. A function $f(x)$ is said to be of *exponential order* if there exist a constant α and positive constants x_0 and M such that

$$|e^{-\alpha x} f(x)| < M$$

for all $x > x_0$ at which $f(x)$ is defined.   □

An example of an exponential order function is $f(x) = x^n$ where $n > 0$. This can be shown as follows. Let $\alpha > 0$ be given. Using L'Hopital's rule n times, we obtain

$$\lim_{x \to \infty} e^{-\alpha x} x^n = 0.$$

Thus, there exist constants $M > 0$ and x_0 such that

$$|e^{-\alpha x} x^n| \le M \quad \text{for all } x \ge x_0.$$

This shows that $f(x) = x^n$ is of exponential order with the constant α any positive number.

In contrast, the function $f(x) = e^{x^2}$ is not of exponential order, since

$$\left| e^{-\alpha x} e^{x^2} \right| = e^{x^2 - \alpha x}$$

can be made larger than any given constant by increasing x.

4.2   Laplace Transform

The Laplace transform of the function $f(x)$ denoted by $\mathcal{L}\{f(x)\}$ is defined by

$$\mathcal{L}\{f(x)\} \equiv \int_0^\infty f(t) e^{-st} dt \equiv F(s), \quad s > 0, \tag{4.3}$$

where $\mathcal{L}$ denotes the Laplace transform operator. The variable s may be complex valued, but in this chapter we restrict s to real numbers.

The Laplace transform operator has a number of properties. For example, $\mathcal{L}$ is a *linear operator*, i.e.,

$$\mathcal{L}\{af(x) \pm bg(x)\} = a\mathcal{L}\{f(x)\} \pm b\mathcal{L}\{g(x)\},$$

where a and b are constants and $f(x)$ and $g(x)$ are functions whose Laplace transforms exist. The second property of the Laplace transform is called the *first shifting property*. It is expressed as

$$\mathcal{L}\{e^{ax} f(x)\} = F(s - a), \quad s > 0. \tag{4.4}$$

The third characteristic is the *second shifting property* which can be expressed as

$$\mathcal{L}\{f(x)\} = e^{-as} F(s). \tag{4.5}$$

Before considering the existence of the Laplace transform, we present some examples.

Example 4.3. Find the Laplace transforms of the following functions:

1. $f(x) = k$,
2. $f(x) = e^{ax}$,

where k and a are constants.

Solution.

1. Here, we have

$$\mathcal{L}\{k\} = \int_0^\infty k e^{-st} dt = \lim_{c \to \infty} \int_0^c k e^{-st} dt = \lim_{c \to \infty} \left[\left(-\frac{k}{s} e^{-st} \right) \Big|_0^c \right] = \frac{k}{s}.$$

2. For the second function $f(x) = e^{ax}$ we compute

$$\mathcal{L}\{e^{ax}\} = \int_0^\infty e^{at} e^{-st} dt = \lim_{c \to \infty} \int_0^c e^{-(s-a)t} dt$$

$$= \lim_{c \to \infty} \left[\left(-\frac{1}{s-a} e^{-(s-a)t} \right) \Big|_0^c \right] = \frac{1}{s-a}, \quad \text{where } s > a.$$

$\square$

Remark 4.4. If we replace a by $-a$ in 2, we obtain $\mathcal{L}\{e^{-ax}\} = \dfrac{1}{s+a}$. $\square$

Remark 4.5. Since $\cosh(ax) = \dfrac{e^{ax} + e^{-ax}}{2}$ and $\sinh(ax) = \dfrac{e^{ax} - e^{-ax}}{2}$, it is easy to show that

- $\mathcal{L}\{\cosh(ax)\} = \dfrac{s}{s^2 - a^2}$,

- $\mathcal{L}\{\sinh(ax)\} = \dfrac{a}{s^2 - a^2}$.

Furthermore, since $\cos(ax) = \dfrac{e^{iax} + e^{-iax}}{2}$ and $\sin(ax) = \dfrac{e^{iax} - e^{-iax}}{2i}$, one can show that

- $\mathcal{L}\{\cos(ax)\} = \dfrac{s}{s^2 + a^2}$,

- $\mathcal{L}\{\sin(ax)\} = \dfrac{a}{s^2 + a^2}$. $\square$

Remark 4.6. Using the first shifting formula (4.4), we obtain

- $\mathcal{L}\{e^{ax}\cos(bx)\} = \dfrac{s - a}{(s - a)^2 + b^2}$,

- $\mathcal{L}\{e^{ax}\sin(bx)\} = \dfrac{b}{(s - a)^2 + b^2}$. $\square$

Example 4.7. Show that $\mathcal{L}\{x^n\} = \dfrac{n!}{s^{n+1}}$, where n is a positive integer.

Solution. We have

$$\mathcal{L}\{x^n\} = \int_0^\infty t^n e^{-st}\, dt = \lim_{c\to\infty} \int_0^c t^n e^{-st}\, dt$$

$$= \lim_{c\to\infty} \left(-\frac{t^n e^{-st}}{s} \bigg|_0^c + \frac{n}{s}\int_0^c t^{n-1} e^{-st}\, dt \right)$$

$$= \cdots = \frac{n(n-1)\cdots 3\cdot 2\cdot 1}{s^n} \lim_{c\to\infty} \int_0^c e^{-st}\, dt = \frac{n!}{s^n}\frac{1}{s} = \frac{n!}{s^{n+1}}.$$

$\square$

Remark 4.8. By Example 4.7, we conclude that if

$$f(x) = a_0 + a_1 x + a_2 x^2 + \cdots,$$

then

$$\mathcal{L}\{f(x)\} = \frac{a_0}{s} + \frac{a_1}{s^2} + 2\frac{a_2}{s^3} + \cdots = \sum_{n=0}^\infty \frac{n!\, a_n}{s^{n+1}}.$$

$\square$

Example 4.9. Find $\mathcal{L}\left\{\dfrac{\sin(x)}{x}\right\}$.

Solution. The series expansion of $\sin(x)$ is given by

$$\sin(x) = \frac{x}{1!} - \frac{x^3}{3!} + \frac{x^5}{5!} - \frac{x^7}{7!} + \cdots$$

Therefore,

$$\mathcal{L}\left\{\frac{\sin(x)}{x}\right\} = \mathcal{L}\left\{\frac{1}{1!} - \frac{x^2}{3!} + \frac{x^4}{5!} - \frac{x^6}{7!} + \cdots\right\} = \frac{\frac{1}{s}}{1!} - \frac{\frac{2!}{s^3}}{3!} + \frac{\frac{4!}{s^5}}{5!} - \frac{\frac{6!}{s^7}}{7!} \cdots$$

$$= \frac{\frac{1}{s}}{1} - \frac{\frac{1}{s^3}}{3} + \frac{\frac{1}{s^5}}{5} - \frac{\frac{1}{s^7}}{7} \cdots = \tan^{-1}\left(\frac{1}{s}\right).$$

Now, one can write

$$\mathcal{L}\left\{\frac{\sin(x)}{x}\right\} = \tan^{-1}\left(\frac{1}{s}\right) = \frac{\pi}{2} - \tan^{-1}(s).$$

□

Example 4.10. A function that plays an important role in statistics is the so-called *error function*

$$\mathrm{erf}(x) = \frac{2}{\sqrt{\pi}} \int_0^x e^{-t^2} dt.$$

Show that $\mathcal{L}\left\{\mathrm{erf}(\sqrt{x})\right\} = \dfrac{1}{s\sqrt{s+1}}$.

Solution. We have

$$\mathcal{L}\left\{\mathrm{erf}\left(\sqrt{x}\right)\right\} = \mathcal{L}\left\{\frac{2}{\sqrt{\pi}} \int_0^{\sqrt{x}} e^{-t^2} dt\right\}$$

$$= \frac{2}{\sqrt{\pi}} \mathcal{L} \int_0^{\sqrt{x}} \left(1 - \frac{t^2}{1!} + \frac{t^4}{2!} - \frac{t^6}{3!} + \cdots\right) dt$$

$$= \frac{2}{\sqrt{\pi}} \mathcal{L}\left\{x^{\frac{1}{2}} - \frac{x^{\frac{3}{2}}}{3 \cdot 1!} + \frac{x^{\frac{5}{2}}}{5 \cdot 2!} - \frac{x^{\frac{7}{2}}}{7 \cdot 3!} + \cdots\right\}$$

$$= \frac{2}{\sqrt{\pi}} \left(\frac{\Gamma\left(\frac{3}{2}\right)}{s^{\frac{3}{2}}} - \frac{\Gamma\left(\frac{5}{2}\right)}{3 \cdot 1! s^{\frac{5}{2}}} + \frac{\Gamma\left(\frac{7}{2}\right)}{5 \cdot 3! s^{\frac{7}{2}}} - \frac{\Gamma\left(\frac{9}{2}\right)}{7 \cdot 5! s^{\frac{9}{2}}} + \cdots\right)$$

$$= \frac{1}{s^{\frac{3}{2}}} - \frac{1}{2 \cdot s^{\frac{5}{2}}} + \frac{1 \cdot 3}{2 \cdot 4 \cdot s^{\frac{7}{2}}} - \frac{1 \cdot 3 \cdot 5}{2 \cdot 4 \cdot 6 \cdot s^{\frac{9}{2}}} + \cdots$$

$$= \frac{1}{s^{\frac{3}{2}}}\left(1 - \frac{1}{2} \cdot \frac{1}{s} + \frac{1 \cdot 3}{2 \cdot 4} \cdot \frac{1}{s^2} - \frac{1 \cdot 3 \cdot 5}{2 \cdot 4 \cdot 6} \cdot \frac{1}{s^3} + \cdots\right)$$

$$= \frac{1}{s^{\frac{3}{2}}}\left(1 + \frac{1}{s}\right)^{-1/2} = \frac{1}{s\sqrt{1+s}}.$$

Here we used the identities $\Gamma(x + 1) = x\Gamma(x)$ and $\Gamma\left(\frac{1}{2}\right) = \sqrt{\pi}$, where x is a positive real number.

$\square$

We come now to the following:

Theorem 4.11. *If the function $f(x)$ is piecewise continuous in any finite interval $0 \le x \le N$ and $|e^{-\alpha x} f(x)| < M$ as $x \to \infty$, then the Laplace transform $L\{f(x)\} = F(s)$ defined by Eq. (4.3) exists for $s > \alpha$.*

Proof. Since $f(x)$ is piecewise continuous, the function $e^{-\alpha x} f(x)$ is integrable on any finite interval of the x-axis. Moreover, $f(x)$ is an exponential order function, i.e., $|f(x)| < e^{\alpha x} M$.

Hence, for $s > \alpha$ it holds

$$|L\{f(x)\}| = \left| \int_0^\infty f(t)e^{-st} dt \right| \le \int_0^\infty |f(t)| e^{-st} dt$$

$$\le M \int_0^\infty e^{\alpha t} e^{-st} dt = \frac{M}{s - \alpha}.$$

This proves the theorem.

$\blacksquare$

It must be noted that the stated conditions are sufficient to guarantee the existence of the Laplace transform. For example, consider $f(x) = \frac{1}{\sqrt{x}}$. This function is not finite in $x = 0$, but its Laplace transform exists, because

$$\left| L\left\{ \frac{1}{\sqrt{x}} \right\} \right| = \int_0^\infty \frac{1}{\sqrt{t}} e^{-st} dt = \frac{1}{\sqrt{s}} \int_0^\infty x^{-\frac{1}{2}} e^{-x} dx = \frac{1}{\sqrt{s}} \Gamma\left(\frac{1}{2}\right) = \sqrt{\frac{\pi}{s}}.$$

Here we used the relation $\Gamma\left(\frac{1}{2}\right) = \sqrt{\pi}$ and in the first integral the change of variables $x \equiv st$.

Example 4.12. Find $L\{u_c(t)\}$ and $L\{P_c(t)\}$, where $u_c(t)$ and $P_c(t)$ are *Heaviside's unit function* and *unit impulse (Dirac delta) function*, respectively, given by

$$u_c(t) = \begin{cases} 1, & t \ge c \\ 0, & t < c \end{cases}, \qquad P_c(t) = \begin{cases} \frac{1}{c}, & 0 \le t \le c \\ 0, & t > c \end{cases}.$$

Solution. We have

$$L\{u_c(t)\} = \int_0^\infty u_c(t)e^{-st} dt = \int_c^\infty e^{-st} dt = \frac{e^{-cs}}{s},$$

$$L\{P_c(t)\} = \int_0^\infty P_c(t)e^{-st} dt = \int_0^c \frac{e^{-st}}{c} dt = \frac{1 - e^{-cs}}{cs}.$$

One can prove that

$$\mathcal{L}\{u_c(t)f(t-c)\} = e^{-cs}F(s), \quad c \geq 0. \tag{4.6}$$

Relation (4.6) can be useful when we deal with a several terms function, in particular if the forcing function in the ODE is a several terms function or a periodic function.

□

Example 4.13. Find the Laplace transform of the function

$$f(t) = \begin{cases} 1, & 0 < t < 1 \\ t, & t > 1 \end{cases}.$$

Solution. Since for $0 < t < 1$ we have $f(t) = 1$, we denote the function by $f_1(t)$. For $t > 1$ we must have $f(t) = t$. Thus, we choose $f_2(t)$ such that for $t \in (0, 1)$ it has the value zero. Clearly, $f_2(t) = tu_1(t)$. But if we write $f(t) = f_1(t) + f_2(t)$, then for $t > 1$, we obtain $f(t) = 1 + t$. Hence, we choose $f_2(t) = (t - 1)u_1(t)$ and obtain

$$f(t) = f_1(t) + f_2(t) = 1 + (t - 1)u_1(t).$$

Now, everything is ready to use Eq. (4.6). We have

$$\mathcal{L}\{f(t)\} = \mathcal{L}\{1\} + \mathcal{L}\{(t - 1)u_1(t)\} = \frac{1}{s} + \frac{e^{-s}}{s^2}.$$

□

Example 4.14. Find the Laplace transform of the function

$$f(t) = \begin{cases} 1, & 0 < t < 1 \\ t, & 1 < t < 2 \\ t^2, & t > 2 \end{cases}.$$

Solution. Again for $0 < t < 1$ we let $f_1(t) = 1$. For $1 < t < 2$, we must have $f(t) = t$. Thus, we choose $f_2(t)$ such that for $t \in (1, 2)$ it has the value zero. Clearly, $f_2(t) = (t - 1)u_1(t)$. Therefore, $f_3(t) = (t^2 - t)u_2(t)$. Then

$$f(t) = f_1(t) + f_2(t) + f_3(t) = 1 + (t - 1)u_1(t) + (t^2 - t)u_2(t).$$

But for using Eq. (4.6), we must have for the third term a function of $(t - 2)$. Hence, we write

$$f(t) = 1 + (t - 1)u_1(t) + [(t - 2)^2 + 3(t - 2) + 2]u_2(t).$$

Now if we apply (4.6), we obtain

$$\mathcal{L}\{f(t)\} = \mathcal{L}\{1\} + \mathcal{L}\{(t-1)u_1(t)\} + \mathcal{L}\{(t-2)^2 u_2(t)\}$$

$$+ 3\mathcal{L}\{(t-2)u_2(t)\} + 2\mathcal{L}\{u_2(t)\}$$

$$= \frac{1}{s} + \frac{e^{-s}}{s^2} + e^{-2s}\left(\frac{2}{s^3} + \frac{3}{s^2} + \frac{2}{s}\right).$$

$\square$

4.3   The Laplace Transform of Periodic Functions

Many problems in applied mathematics deal with periodic functions. Most inhomogeneous ODEs in engineering involve forcing functions that are periodic. We wish to find the Laplace transform of such functions.

A periodic function $f(x)$ is one that is piecewise continuous and for some $p > 0$ satisfies

$$f(x) = f(x+p) = f(x+2p) = \cdots = f(x+np) = \cdots$$

Using this relation, we can write the transform of $f(x)$ as a series of integrals

$$\mathcal{L}\{f(x)\} = \int_0^\infty f(t)e^{-st}\,dt = \int_0^p f(t)e^{-st}\,dt + \int_p^{2p} f(t)e^{-st}\,dt + \cdots$$

Now if in the first integral we let $t = u$ and in second integral we let $t = u + p$, and in third one $t = u + 3p$ etc., we obtain

$$\mathcal{L}\{f(x)\}$$

$$= \int_0^p f(u)e^{-su}\,du + \int_0^p f(u)e^{-s(u+p)}\,du + \int_0^p f(u)e^{-s(u+2p)}\,du + \cdots$$

$$= \int_0^p f(u)e^{-su}\,du + e^{-sp}\int_0^p f(u)e^{-su}\,du + e^{-2sp}\int_0^p f(u)e^{-su}\,du + \cdots$$

$$= \left(1 + e^{-sp} + e^{-2sp} + \cdots\right)\int_0^p f(u)e^{-su}\,du$$

$$= (1 - e^{-sp})^{-1}\int_0^p f(u)e^{-su}\,du.$$

Therefore, we have proved

Theorem 4.15. *If* $f(x)$ *is piecewise continuous with period* $p > 0$, *then*

$$\mathcal{L}\{f(x)\} = (1 - e^{-sp})^{-1} \int_0^p f(u)e^{-su}du. \tag{4.7}$$

Example 4.16. Find the Laplace transform of the half-way rectification given by

$$f(x) = \begin{cases} \sin(x), & 0 < x < \pi \\ 0, & \pi < x < 2\pi \end{cases}, \quad p = 2\pi.$$

Solution. Using Eq. (4.7), we obtain

$$\mathcal{L}\{f(x)\}$$

$$= \left(1 - e^{-2\pi s}\right)^{-1} \int_0^{2\pi} f(u)e^{-su}du = \left(1 - e^{-2\pi s}\right)^{-1} \int_0^{\pi} \sin(u)e^{-su}du$$

$$= \frac{1 - e^{-\pi s}}{(1 - e^{-2\pi s})(1 + s^2)} = \frac{1}{(1 + e^{-\pi s})(1 + s^2)}.$$

□

Example 4.17. Find the Laplace transform of the wave function with period $2p$.

Solution. We have

$$f(x) = \begin{cases} k, & 0 < x < p \\ -k, & p < x < 2p. \end{cases}$$

Hence,

$$\mathcal{L}\{f(x)\} = \left(1 - e^{-2ps}\right)^{-1} \int_0^{2p} f(u)e^{-su}du$$

$$= \left(1 - e^{-2ps}\right)^{-1} \left[\int_0^p ke^{-su}du - \int_p^{2p} ke^{-su}du \right]$$

$$= \frac{k(1 - e^{-ps})}{s(1 + e^{-ps})}.$$

□

Example 4.18. Find the Laplace transform of the saw-tooth wave function defined by

$$f(x) = \frac{x}{p} \quad \text{when } 0 < x < p \quad \text{and} \quad f(x + p) = f(x).$$

Solution. We have

$$\mathcal{L}\{f(x)\} = \frac{1}{p(1 - e^{-ps})} \int_0^p u e^{-su} du$$

$$= \frac{1}{p(1 - e^{-ps})} \left[\frac{u e^{-su}}{s} \right]_0^p - \frac{1}{s} \int_0^p e^{-su} du$$

$$= \frac{1 - (1 + sp)e^{-ps}}{ps^2(1 - e^{-ps})}.$$

□

Definition 4.19. Let us suppose that there exists a function $f(x)$ such that $\mathcal{L}\{f(x)\} = F(s)$. Then $f(x)$ is called the *inverse Laplace transform* of $F(s)$ and will be denoted by $\mathcal{L}^{-1}$, i.e. $f(x) = \mathcal{L}^{-1}\{F(s)\}$. □

It must be noted that for finding the inverse Laplace transform, in most problems, we may factoring the denominator and expressing the function in terms of its partial fraction expansion.

Example 4.20. Find the inverse of the transformed function

$$F(s) = \frac{2s - 3}{s^2 - 3s - 4}.$$

Solution. We have

$$\mathcal{L}^{-1}\{F(s)\} = \mathcal{L}^{-1}\left\{ \frac{2s - 3}{s^2 - 3s - 4} \right\} = \mathcal{L}^{-1}\left\{ \frac{1}{s + 1} + \frac{1}{s - 4} \right\}$$

$$= \mathcal{L}^{-1}\left\{ \frac{1}{s + 1} \right\} + \mathcal{L}^{-1}\left\{ \frac{1}{s - 4} \right\} = e^{-x} + e^{4x}.$$

□

Example 4.21. Find $\mathcal{L}^{-1}\left\{ \frac{2s}{(s^2 + 1)(s^2 + 4)} \right\}$.

Solution. Employing partial fractions, we get

$$\frac{2s}{(s^2 + 1)(s^2 + 4)} = \frac{As + B}{s^2 + 1} + \frac{Cs + D}{s^2 + 4}$$

$$= \frac{(As + B)(s^2 + 4) + (Cs + D)(s^2 + 1)}{(s^2 + 1)(s^2 + 4)},$$

Thus we obtain

$$2s = (As + B)(s^2 + 4) + (Cs + D)(s^2 + 1).$$

Ordering the terms in powers of s and comparing both sides it is an easy but tedious task to show that

$$A = -B = \frac{2}{3}, \quad C = D = 0.$$

Hence

$$\mathcal{L}^{-1}\left\{\frac{2s}{(s^2 + 1)(s^2 + 4)}\right\} = \frac{2}{3}\mathcal{L}^{-1}\left\{\frac{s}{s^2 + 1}\right\} - \frac{2}{3}\mathcal{L}^{-1}\left\{\frac{s}{s^2 + 4}\right\}$$

$$= \frac{2}{3}(\cos(x) - \cos(2x)).$$

In general this process of finding the constants takes time. Later, by using the convolution, we try to reduce this task.

□

4.4 The Laplace Transforms of Derivatives and Integrals

For $x \geq 0$, let the function $f(x)$ be piecewise continuous and of exponential order. However, the exponential order of the function $f(x)$ does not imply that its derivatives are also of exponential order. For example, $f(x) = \sin(e^{x^2})$ is of exponential order but its derivative has not this property. By imposing stronger conditions on the function as in Theorem 4.11, one can prove the following result.

Theorem 4.22. *Assume that for $x \geq 0$ the function $f(x)$ is continuous and $f'(x)$ is piecewise continuous. Furthermore, let f and f' be of exponential order when $x \to \infty$. Then*

$$\mathcal{L}\{f'(x)\} = s\mathcal{L}\{f(x)\} - f(0). \tag{4.8}$$

In a similar way, if f and f' are continuous functions and in Theorem 4.22 the conditions posed on f' are applied to f'', it holds

$$\mathcal{L}\{f''(x)\} = s^2\mathcal{L}\{f(x)\} - sf(0) - f'(0). \tag{4.9}$$

These results can be extended to higher-order derivatives and give us

$$\mathcal{L}\{f^{(n)}(x)\} = s^n\mathcal{L}\{f(x)\} - s^{n-1}f(0) - s^{n-2}f'(0) - \cdots - f^{(n-1)}(0). \tag{4.10}$$

Equation (4.10) represents a fundamental relation for the application of the Laplace transform to solve an ODE with constant coefficients. The differential equation is transformed into an algebraic expression in terms of s, and the solution of the ODE itself will result from the determination of the inverse Laplace transform which we discussed before.

Example 4.23. Solve the initial value problem given by

$$x''(t) + \alpha^2 x(t) = A\sin(\omega t), \quad x(0) = x'(0) = 0.$$

Solution. This initial value problem describes the forced, undamped, and resonant motion of a mass on a spring such that $\alpha^2 = k/m$, where k is the spring constant. The mass starts from rest. We want to find the displacement x in terms of t. Under the given initial conditions, we apply the Laplace transform to both sides of this equation. This yields

$$X(t) = \frac{A\omega}{(s^2 + \omega^2)(s^2 + \alpha^2)}.$$

Now, since $X(t) = \mathcal{L}\{x(t)\}$, the inverse Laplace transform gives

$$x(t) = \mathcal{L}^{-1}\left\{\frac{A\omega}{(s^2 + \omega^2)(s^2 + \alpha^2)}\right\} = \frac{A}{\alpha^2 - \omega^2}\mathcal{L}^{-1}\left\{\frac{\omega}{s^2 + \omega^2} - \frac{\omega}{s^2 + \alpha^2}\right\}$$

$$= \frac{A}{\alpha^2 - \omega^2}\left(\sin(\omega t) - \frac{\omega}{\alpha}\sin(\alpha t)\right).$$

$\square$

Example 4.24. Solve the initial value problem

$$x''(t) + 4x(t) = f(t), \quad x(0) = x'(0) = 0,$$

where $f(t) = \begin{cases} \sin(2t), & 0 \le t < \pi \\ 0, & t \ge \pi \end{cases}$.

Solution. Here the forcing function can be interpreted as an external force which is acting on a mechanical system only for a short time and is then removed. We write

$$x''(t) + 4x(t) = \sin(2t) - u_\pi \sin(2(t - \pi)).$$

Taking the Laplace transform and implying the initial conditions, we obtain

$$\mathcal{L}\{x(t)\} = \frac{2}{(s^2 + 4)^2} - \frac{2e^{-\pi s}}{(s^2 + 4)^2}.$$

Hence

$$x(t) = \mathcal{L}^{-1} \left\{ \frac{2}{(s^2 + 4)^2} \right\} - \mathcal{L}^{-1} \left\{ \frac{2e^{-\pi s}}{(s^2 + 4)^2} \right\}$$

$$= \frac{\sin(2t) - 2t \cos(2t)}{8} - \frac{(t - \pi) \sin(2(t - \pi))}{4} u_\pi.$$

□

Example 4.25. Find the response of an *RL* circuit to a unit pulse function $u_a - u_b$. Assume that the initial current is zero.

Solution. We know that the initial value problem describing this circuit is

$$Li' + Ri = u_a - u_b, \quad i(0) = 0.$$

Taking the Laplace transform, we obtain

$$LsI(s) + RI(s) = \frac{e^{-as}}{s} - \frac{e^{-bs}}{s},$$

where $I(s) = \mathcal{L}\{i(t)\}$. Writing

$$I(s) = \frac{e^{-as} - e^{-bs}}{s(Ls + R)}$$

and factoring the denominator to express the right-hand side in terms of its partial fraction expansion, we get

$$I(s) = \frac{e^{-as}}{R} \left(\frac{1}{s} - \frac{1}{s + R/L} \right) - \frac{e^{-bs}}{R} \left(\frac{1}{s} - \frac{1}{s + R/L} \right).$$

Taking the inverse Laplace transform, we obtain

$$i(t) = \frac{1}{R} \left[\left(1 - e^{-\frac{R(t-a)}{L}}\right) u_a - \left(1 - e^{-\frac{R(t-b)}{L}}\right) u_b \right].$$

□

Example 4.26. Let a beam whose ends are at $x = 0$ and $x = l$ is coincident with the x-axis and a vertical load, given by $W(x)$ per unit length, acts transversely on it. Then the axis of the beam has a transverse deflection $y(x)$ at the point x which satisfies the ODE

$$EIy^{(4)}(x) = W(x), \quad 0 < x < l,$$

where EI is assumed to be constant. The boundary conditions associated with this equation depend on the manner in which the beam is supported. For example, for a free end, we have $y'' = y''' = 0$; for a hinged or supported end, $y = y'' = 0$; and for a clamped, built-in, or fixed end, $y = y' = 0$. Let $l = 2$. If the beam is clamped at $x = 0$ and is free at the other end, find the deflection if it carries a load given by

$$W(x) = \begin{cases} 8, & 0 < x < 1 \\ 0, & 1 < x < 2 \end{cases}.$$

Solution. We have

$$EI y^{(4)}(x) = \begin{cases} 8, & 0 < x < 1 \\ 0, & 1 < x < 2 \end{cases},$$

with the initial conditions $y(0) = y'(0) = 0$, $y''(2) = y'''(2) = 0$. We write the right-hand side in terms of the unit function. Thus, we get

$$EI y^{(4)}(x) = 8(u_0 - u_1).$$

Taking the Laplace transform and implying the initial conditions, we obtain

$$s^4 \mathcal{L}\{y(x)\} - sy''(0) - y'''(0) = \frac{8}{EI}\left(\frac{1 - e^{-s}}{s}\right).$$

Replacing the two unknown initial conditions by $y''(0) = A$, $y'''(0) = B$, we get

$$\mathcal{L}\{y(x)\} = \frac{A}{s^3} + \frac{B}{s^4} + \frac{8}{EI}\left(\frac{1 - e^{-s}}{s^5}\right).$$

Thus,

$$y(x) = \mathcal{L}^{-1}\left\{\frac{A}{s^3}\right\} + \mathcal{L}^{-1}\left\{\frac{B}{s^4}\right\} + \frac{8}{EI}\mathcal{L}^{-1}\left\{\frac{1 - e^{-s}}{s^5}\right\}$$

$$= \frac{Ax^2}{2} + \frac{Bx^3}{6} + \frac{u_1}{3EI}(x - 1)^4.$$

Using $y''(2) = y'''(2) = 0$, one can show that $A = \dfrac{1}{EI}$ and $B = \dfrac{-4}{EI}$.

$\square$

We may now turn our attention to the integration process, and we expect to have a division of the transform by s since the integration of $f(x)$ is the inverse operation of the differentiation. We continue the present section with the following theorem.

Theorem 4.27. *If the function $f(x)$ satisfies the assumptions of Theorem 4.11, then*

$$\mathcal{L}\left\{\int_0^x f(t)dt\right\} = \frac{F(s)}{s}. \tag{4.11}$$

Proof. Let $g(x) \equiv \int_0^x f(t)dt$. Thus, $g'(x) = f(x)$. Since $g(0) = 0$, we have

$$\mathcal{L}\{g'(x)\} = s\mathcal{L}\{g(x)\} - g(0) = s\mathcal{L}\{g(x)\}.$$

Hence

$$\mathcal{L}\{g(x)\} = \frac{1}{s}\mathcal{L}\{g'(x)\} = \frac{1}{s}\mathcal{L}\{f(x)\} = \frac{F(s)}{s}.$$

■

Example 4.28. Find $\mathcal{L}\left\{\int_0^x \frac{\sin(t)}{t}dt\right\}$.

Solution. Since $\mathcal{L}\left\{\frac{\sin(x)}{x}\right\} = \tan^{-1}\left(\frac{1}{s}\right)$, we have

$$\mathcal{L}\left\{\int_0^x \frac{\sin(t)}{t}dt\right\} = \frac{1}{s}\tan^{-1}\left(\frac{1}{s}\right).$$

□

Remark 4.29. Looking at (4.11), one may write

$$\mathcal{L}^{-1}\left\{\frac{F(s)}{s}\right\} = \int_0^x f(t)dt. \tag{4.12}$$

Equation (4.12) can be used to find $f(x)$.   □

Example 4.30. If $f(x) = \mathcal{L}^{-1}\left\{\frac{1}{s(s^2 + a^2)}\right\}$, find $f(x)$.

Solution. Since

$$\mathcal{L}^{-1}\left\{\frac{1}{s^2 + a^2}\right\} = \frac{\sin(ax)}{a}$$

it holds

$$f(x) = \mathcal{L}^{-1}\left\{\frac{1}{s(s^2 + a^2)}\right\} = \int_0^x \frac{\sin(at)}{a}dt = \frac{1 - \cos(ax)}{a^2}.$$

□

4.5 Derivatives and Integrals of the Laplace Transforms

We described the Laplace transform of derivatives and integrals. In advanced calculus, it can be shown that differentiation and integration of the Laplace transform
are possible. For example, if $F(s) = \int_0^\infty f(t)e^{-st}\,dt$, then using Leibniz's rule of
differentiation and integration we have

$$F'(s) = -\int_0^\infty t f(t)e^{-st}\,dt = \int_0^\infty (-t f(t))e^{-st}\,dt = \mathcal{L}\{-x f(x)\}\,.$$

That is, a Laplace transform corresponds to $f(x)$ times $(-x)$. By the same manner,
the second derivative is

$$F''(s) = \int_0^\infty t^2 f(t)e^{-st}\,dt = \int_0^\infty (t^2 f(t))e^{-st}\,dt = \mathcal{L}\{x^2 f(x)\}\,.$$

These operations enable us to formulate the following theorem.

Theorem 4.31. *If $f(x)$ and its derivatives satisfy the conditions of Theorem 4.11,
then*

$$F^{(n)}(s) = \mathcal{L}\{(-x)^n f(x)\}\,.$$

This equation can be rewritten as

$$\mathcal{L}\{x^n f(x)\} = (-1)^n F^{(n)}(s). \tag{4.13}$$

Example 4.32. Find $\mathcal{L}\{x^2 \sin(3x)\}$.

Solution. We have

$$\mathcal{L}\{x^2 \sin(3x)\} = \frac{d^2}{ds^2}\left(\frac{3}{s^2+9}\right) = \frac{6s(3s^2-9)}{(s^2+9)^3}\,.$$

□

The relation (4.13) can also be used to evaluate some improper integrals.

Example 4.33. Show that $\displaystyle\int_0^\infty x^2 \cos(x)e^{-x}\,dx = -\frac{1}{2}$.

Solution. By relation (4.13), we have

$$\int_0^\infty x^2 \cos(x)e^{-sx}\,dx = \mathcal{L}\{x^2 \cos(x)\} = \frac{d^2}{ds^2}\mathcal{L}\{\cos(x)\}$$

$$= \frac{d^2}{ds^2}\left(\frac{s}{s^2+1}\right) = \frac{2s(s^2-3)}{(s^2+1)^3}\,.$$

Now set $s = 1$ to get the desired result.

□

Example 4.34. Let $\mathcal{L}\{f(x)\} = \ln\left(\dfrac{s+1}{s+2}\right) = F(s)$. Find $f(x)$.

Solution. We write $\ln\left(\dfrac{s+1}{s+2}\right) = \ln(s+1) - \ln(s+2)$. Thus,

$$F'(s) = \frac{1}{s+1} - \frac{1}{s+2} = \mathcal{L}\{e^{-x}\} - \mathcal{L}\{e^{-2x}\} = \mathcal{L}\{e^{-x} - e^{-2x}\}.$$

Now, if we use the relation (4.13) with $n = 1$, we obtain

$$xf(x) = e^{-2x} - e^{-x}, \quad \text{i.e. } f(x) = \frac{e^{-2x} - e^{-x}}{x}.$$

Suppose $g(x) \equiv f(x)/x$, then $f(x) = xg(x)$. But we have

$$\mathcal{L}\{xg(x)\} = -G'(s).$$

Thus, $F(s) = -G'(s)$. That is, $-dG = F(s)ds$. Integrating both sides, we get

$$-G(s) = \int_\infty^s F(t)dt,$$

where $G(s) \to 0$ as $s \to \infty$. Hence $G(s) = \displaystyle\int_s^\infty F(t)dt$. Thus,

$$\mathcal{L}\left\{\frac{f(x)}{x}\right\} = \int_s^\infty F(t)dt, \tag{4.14}$$

provided $\displaystyle\lim_{x\to 0^+} \frac{f(x)}{x}$ exists.

□

Example 4.35. Find $\mathcal{L}\left\{\dfrac{e^{-4x} - 1}{x}\right\}$.

Solution. Since $\displaystyle\lim_{x\to 0^+}\left(\dfrac{e^{-4x} - 1}{x}\right) = 1$ and $\mathcal{L}\{e^{-4x} - 1\} = \dfrac{1}{s+4} - \dfrac{1}{s}$, we have

$$\mathcal{L}\left\{\frac{e^{-4x} - 1}{x}\right\} = \int_s^\infty \left(\frac{1}{t+4} - \frac{1}{t}\right)dt = \ln\left(\frac{s}{s+1}\right).$$

□

Example 4.36. Evaluate $\int_0^\infty \dfrac{e^{-2x}\sin(x)\cosh(x)}{x}dx.$

Solution. We have $\lim\limits_{x\to 0+} \dfrac{\sin(x)\cosh(x)}{x} = 1$, hence

$$\mathcal{L}\left\{\frac{\sin(x)\cosh(x)}{x}\right\} = \int_0^\infty \frac{e^{-st}\sin(t)\cosh(t)}{t}dt = F(s).$$

But

$$\mathcal{L}\left\{\frac{\sin(x)\cosh(x)}{x}\right\} = \int_s^\infty \mathcal{L}\{\sin(t)\cosh(t)\}\,dt = F(s).$$

Comparing these results, we obtain

$$\int_0^\infty \frac{e^{-st}\sin(t)\cosh(t)}{t}dt = \int_s^\infty \mathcal{L}\{\sin(t)\cosh(t)\}\,dt$$

$$= \frac{1}{2}\int_s^\infty \left[\mathcal{L}\{e^t\sin(t)\} + \mathcal{L}\{e^{-t}\sin(t)\}\right]dt$$

$$= \frac{1}{2}\int_s^\infty \left[\frac{1}{(t-1)^2+1} + \frac{1}{(t+1)^2+1}\right]dt$$

$$= \frac{1}{2}\left[\tan^{-1}(s-1) + \tan^{-1}(s+1)\right].$$

Now, if we set $s = 2$, we get

$$\int_0^\infty \frac{e^{-2x}\sin(x)\cosh(x)}{x}dx = \frac{1}{2}\left(\frac{\pi}{4} + \tan^{-1}(3)\right).$$

□

Example 4.37. Solve the initial value problem given by

$$xy'' + 2y' + xy = 0, \quad y(0^+) = 1, \ y(\pi) = 0.$$

Solution. By taking the Laplace transform of the given equation, we obtain

$$\mathcal{L}\{xy''\} + 2\mathcal{L}\{y'\} + \mathcal{L}\{xy\}$$

$$= -\left[s^2Y - sy(0^+) - y'(0^+)\right]' + 2\left[sY - y(0^+)\right] - Y' = 0,$$

where $Y \equiv \mathcal{L}\{y\}$. Since $y(0^+) = 1$, we have

$$Y' = -\frac{1}{s^2+1}, \quad \text{i.e.} \quad Y(s) = c - \tan^{-1}(s).$$

But, $Y \to 0$ as $s \to \infty$, hence $c = \dfrac{\pi}{2}$. Thus,

$$Y(s) = \frac{\pi}{2} - \tan^{-1}(s) = \tan^{-1}\left(\frac{1}{s}\right).$$

Since $\mathcal{L}\left\{\dfrac{\sin(x)}{x}\right\} = \tan^{-1}\left(\dfrac{1}{s}\right)$, we obtain

$$\mathcal{L}^{-1}\{Y\} = \mathcal{L}^{-1}\left\{\tan^{-1}\left(\frac{1}{s}\right)\right\} \to y(x) = \frac{\sin(x)}{x}.$$

This solution satisfies $y(\pi) = 0$.

$\square$

4.6   The Convolution Theorem and Integral Equations

As we observed, for obtaining the inverse Laplace transform of $F(s)$, the partial fraction expansion can be applied. In general, such an expansion is not difficult, but it may be tedious to develop it. In this section, we introduce the concept of the *convolution* which can be used not only for finding the inverse of the product of two transform functions but also to solve some *integral equations* that are of the form of a convolution integral.

Definition 4.38. Suppose $f(x)$ and $g(x)$ are piecewise continuous functions. The convolution $f * g$ of the functions f and g is defined by

$$f * g \equiv \int_0^x f(t)g(x-t)dt.$$

$\square$

The convolution $f * g$ has the following properties:

- $f * g = g * f$ (commutative law),
- $f * (g * h) = (f * g) * h$ (associative law),
- $f * (g + h) = f * g + f * h$ (distributive law).

The next theorem contains a fundamental statement which is concerned with the transform of a convolution function.

Theorem 4.39. *If* $\mathcal{L}\{f(x)\} = F(s)$ *and* $\mathcal{L}\{g(x)\} = G(s)$, *then*

$$\mathcal{L}^{-1}\{F(s)G(s)\} = \int_0^x f(t)g(x-t)dt. \tag{4.15}$$

Proof. We have

$$\mathcal{L}\left\{\int_0^x f(t)g(x-t)dt\right\} = \int_0^\infty e^{-sx}\left\{\int_0^x f(t)g(x-t)dt\right\}dx.$$

It follows

$$\mathcal{L}\left\{\int_0^x f(t)g(x-t)dt\right\} = \int_0^\infty \int_0^x e^{-sx} f(t)g(x-t)dt\, dx. \qquad (4.16)$$

Now, if we assume that $x = u + t$, then (4.16) becomes

$$\mathcal{L}\left\{\int_0^x f(t)g(x-t)dt\right\} = \int_0^\infty \int_0^\infty e^{-s(u+t)} f(t)g(u)dt\, du.$$

The double integral of the right-hand side can be written as the product of two integrals. Then

$$\mathcal{L}\left\{\int_0^x f(t)g(x-t)dt\right\} = \int_0^\infty e^{-st} f(t)dt \int_0^\infty e^{-su} g(u)du$$

$$= \mathcal{L}\{f(x)\}\mathcal{L}\{g(x)\}.$$

But $\mathcal{L}\{f(x)\}\mathcal{L}\{g(x)\} = F(s)G(s)$. Thus,

$$\mathcal{L}\left\{\int_0^x f(t)g(x-t)dt\right\} = F(s)G(s),$$

and the proof is completed.   ■

Let us return to Example 4.21. Our aim was to find

$$\mathcal{L}^{-1}\left\{\frac{2s}{(s^2+1)(s^2+4)}\right\}.$$

Hence we write

$$\mathcal{L}^{-1}\left\{\frac{2s}{(s^2+1)(s^2+4)}\right\} = \mathcal{L}^{-1}\left\{\frac{s}{s^2+1} \times \frac{2}{s^2+4}\right\}$$

and use Theorem 4.39. The result is

$$\mathcal{L}^{-1}\left\{\frac{s}{s^2+1} \times \frac{2}{s^2+4}\right\} = \int_0^x \cos(x-t)\sin(2t)dt = \frac{2}{3}(\cos(x) - \cos(2x)).$$

Example 4.40. Find $\mathcal{L}^{-1}\left\{\frac{1}{(s^2+9)s^5}\right\}.$

Solution. We can write

$$\mathcal{L}^{-1}\left\{\frac{1}{(s^2+9)s^3}\right\} = \frac{1}{6}\mathcal{L}^{-1}\left\{\frac{3}{s^2+9} \times \frac{2}{s^3}\right\} = \frac{1}{6}\int_0^x (x-t)^2 \sin(3t)dt.$$

Integrating by parts leads to

$$\mathcal{L}^{-1}\left\{\frac{1}{(s^2+9)s^3}\right\} = \frac{x^2}{18} + \frac{1}{81}(\cos(3x)-1).$$

□

As we mentioned before, Theorem 4.39 may be applied to solve some *integral equations* that are of the form of a convolution integral.

Consider the integral equation given by

$$f(x) = h(x) + \lambda \int_0^x f(t)g(x-t)dt. \tag{4.17}$$

Taking the Laplace transform, we obtain

$$F(s) = H(s) + \lambda F(s)G(s),$$

where $F(s) = \mathcal{L}\{f(x)\}$, $G(s) = \mathcal{L}\{g(x)\}$, $H(s) = \mathcal{L}\{h(x)\}$, and λ is a constant. We can write

$$F(s) = \frac{H(s)}{1 + \lambda G(s)}.$$

Now, take the inverse Laplace transform to find $f(x)$.

Example 4.41. Solve the integral equation given by

$$f(x) = x + \int_0^x f(t)e^{x-t}dt.$$

Solution. We have $G(s) = \mathcal{L}\{e^x\} = \dfrac{1}{s-1}$, $H(s) = \mathcal{L}\{x\} = \dfrac{1}{s^2}$ and $\lambda = 1$. Thus,

$$F(s) = \frac{H(s)}{1 + \lambda G(s)} = \frac{s-1}{s^3} = \frac{1}{s^2} - \frac{1}{s^3}.$$

It follows

$$f(x) = \mathcal{L}^{-1}\left\{\frac{1}{s^2}\right\} - \mathcal{L}^{-1}\left\{\frac{1}{s^3}\right\} = x - \frac{x^2}{2}.$$

□

Example 4.42. Solve the integral equation given by

$$f'(x) = \sin(x) + \int_0^x f(x - t)\cos(t)dt, \quad f'(0) = 0.$$

Solution. Taking the Laplace transform and using the initial condition, we get

$$sF(s) = \frac{1}{s^2 + 1} + \frac{sF(s)}{s^2 + 1}, \quad \text{i.e.} \quad F(s) = \frac{1}{s^3}.$$

From this we obtain

$$f(x) = \frac{x^2}{2}.$$

□

The equation in Example 4.42 is a so-called integro-differential equation. That is, the Laplace transform can also be used to solve such equations. In the next chapter, we use the Laplace transform to solve two or more simultaneous ODEs.

4.7   Exercises

Exercise 4.1. Show that:

(1) $\mathcal{L}\left\{f\left(\frac{a}{x}\right)e^{-\frac{bx}{a}}\right\} = F(as + b)$,

(2) $\mathcal{L}\{x^n e^{ax}\} = \dfrac{n!}{(s - a)^{n+1}}$,

(3) $\mathcal{L}\{e^{-ax}\sin(bx)\} = \dfrac{n!}{(s - a)^{n+1}}$,

(4) $\mathcal{L}\left\{\dfrac{1 - \cos(ax)}{x}\right\} = \dfrac{1}{2}\ln\left(1 + \dfrac{a^2}{s^2}\right)$,

(5) $\mathcal{L}\{\sqrt{x}\} = \dfrac{1}{2s}\sqrt{\dfrac{\pi}{s}}$,

(6) $\mathcal{L}\{\sin(\sqrt{x})\} = \dfrac{e^{-1/4s}}{2s}\sqrt{\dfrac{\pi}{s}}$,

(7) $\mathcal{L}\left\{\dfrac{\cos(\sqrt{x})}{\sqrt{x}}\right\} = e^{-1/4s}\sqrt{\dfrac{\pi}{s}}$,

(8) $\mathcal{L}\left\{\int_0^\infty \frac{f(t)x^t}{t!}dt\right\} = \frac{1}{2s}\sqrt{\frac{\pi}{s}}$,

(9) $\mathcal{L}\{f(x^2)\} = \frac{1}{2\sqrt{\pi}}\int_0^\infty \frac{f(t)e^{-s^2/4t}}{t\sqrt{t}}dt$,

(10) $\mathcal{L}\left\{\frac{1}{\sqrt{\pi x}}\int_0^\infty f(t)e^{-t^2/4x}dt\right\} = \frac{F(\ln(s))}{\sin(s)}$.

Exercise 4.2. Determine the solution of the following initial value problems:

(1) $y'' - 3y' + 2y = 4\cos(x)$, $y(0) = 0,\ y'(0) = 1$,

(2) $y'' + 4y' + 3y = x + 4$, $y(0) = 1,\ y'(0) = 0$,

(3) $y''' - 3y'' - 4y' = 2e^x - \sin(x)$, $y(0) = y'(0) = y'''(0) = 0$,

(4) $y'' + 4y = \sin(x) - \sin(x - 2\pi)u_{2\pi}$, $y(0) = y'(0) = 0$,

(5) $y'' + y = (x - 2)u_2$, $y(0) = y'(0) = 0$,

(6) $xy'' + (3x - 1)y' - (4x - 9)y = 0$, $y(0) = y'(0) = 0$,

(7) $xy'' + (x - 1)y' - y = 0$, $y(0) = 5,\ y'(\infty) = 0$,

(8) $y'' - xy' + y = 1$, $y(0) = 1,\ y'(0) = -2$,

(9) $y'' + y = \begin{cases} 4, & 0 \le x \le 2 \\ x + 2, & x > 2 \end{cases}$, $y(0) = 1,\ y'(0) = 0$,

(10) $y'' + 4y = \begin{cases} x, & 0 \le x \le 1 \\ 1, & x > 1 \end{cases}$, $y(0) = 1,\ y'(0) = 0$.

Exercise 4.3. Find the inverse Laplace transform of following expressions:

(1) $\dfrac{s^2}{(s^2 + 1)(s^2 + 4^2)}$,

(2) $\dfrac{s^2}{(s^2 + 9)^4}$,

(3) $e^{-2s}\ln\left(\dfrac{s}{s-1}\right)$,

(4) $\dfrac{1}{s}\ln\left(\dfrac{s^2 + a^2}{s^2 + b^2}\right)$,

(5) $\dfrac{1 - se^{-\pi s}}{s^2 + 1}$,

(6) $\dfrac{1}{s^3 + 1}$,

(7) $\dfrac{3(s^2 + 1)e^{-4s}}{s^6}$,

(8) $\dfrac{(1 - e^{-s})e^{-s}}{s(s^2 + 1)}$,

(9) $e^{-as}\tan^{-1}(s - 1),\ a > 0$,

(10) $\dfrac{s}{s^2 + 1}\cot^{-1}(s + 1)$.

Exercise 4.4. Verify that:

(1) $\displaystyle\int_0^\infty \frac{\cos(ax)-\cos(bx)}{x}dx = \ln\left(\frac{a}{b}\right),$ (2) $\displaystyle\int_0^\infty \frac{\cos(ax)}{1+x^2}dx = \frac{\pi}{2}e^{-a}.$

(3) $\displaystyle\int_0^\infty \frac{x\sin(ax)}{1+x^2}dx = \frac{\pi}{2}e^{-a},\ a>0,$ (4) $\displaystyle\int_0^\infty \frac{\sin(ax)}{x}dx = \frac{\pi}{2},$

(5) $\displaystyle\int_0^\infty \frac{e^{-ax}\sin(bx)}{x}dx = \cot^{-1}\left(\frac{b}{a}\right),$ (6) $\displaystyle\int_0^\infty \sin(x^2)dx = \frac{1}{2}\sqrt{\frac{\pi}{2}},$

(7) $\displaystyle\int_0^\infty x^3 e^{-x}\sin(x)dx = 0,$ (8) $\displaystyle\int_0^\infty xe^{\pm 2x}\cos(x)dx = 0,$

(9) $\displaystyle\int_0^\infty \frac{e^{-ax}-e^{-bx}}{x}dx = \ln\left(\frac{b}{a}\right),$ (10) $\displaystyle\int_0^\infty x^2\cos(x)dx = 0,$

Exercise 4.5. Evaluate the following integrals:

(1) $\displaystyle\int_0^\infty x^3 e^{-x}\cos(x)dx,$ (2) $\displaystyle\int_0^\infty (-\ln(x))^{\frac{1}{2}}dx,$

(3) $\displaystyle\mathcal{L}\left\{\frac{1}{\pi}\int_0^\infty \cos(x\cos(x))dx\right\},$ (4) $\displaystyle\int_0^\infty xe^{-x}\mathrm{erf}(x)dx,$

(5) $\displaystyle\int_0^\infty xe^{-3x}\left(\sin^2(x)\right)\left[\mathcal{L}^{-1}\left\{\ln\left(\frac{s}{s-1}\right)\right\}\right]dx.$

Exercise 4.6. Solve the following integral and integro-differential equations:

(1) $\displaystyle f(x) = x^2 + \int_0^x f(x-t)\sin(t)dt,$

(2) $\displaystyle f(x) = \cos(x) + \int_0^x f(t)e^{x-t}dt,$

(3) $\displaystyle f(x) = \cos(x) + e^{2x}\int_0^x f(t)e^{-2t}dt + \int_0^x f(x-t)e^{-2t}dt,$

(4) $\displaystyle y'(x) = x + \int_0^x y(x-t)\cos(t)dt,\quad y'(0) = 0,$

(5) $\displaystyle y'(x) + 2y - 3\int_0^x y(t)dt = 5(1+x),\quad y'(0) = 2,$

(6) $\displaystyle y''(x) = \cos(3x) + \int_0^x (x-t)y'(t)dt,\quad y(0) = y'(0) = 0,$

(7) $y''(x) + y'(x) = \int_0^x \sin(x - t)y'(t)dt, \quad y(0) = y'(0) = 1,$

(8) $y''(x) + y'(x) = \cos(x) + \int_0^x \sin(t)y'(x - t)dt, \quad y(0) = y'(0) = 0,$

(9) $y(x) = \sin(x) - \int_0^x ty(x - t)\mathcal{L}^{-1} \{\cot^{-1}(s)\} dt, \quad y(0) = y'(0) = 1,$

(10) $y''(x) - y'(x) = \int_0^x y''(t)y'(x - t)dt, \quad y(0) = y'(0) = 0.$

Exercise 4.7. The differential equation

$$y''(t) + 5y(t) = \frac{1}{2}\delta(t - 2)$$

describes the equation of motion for a mechanical system where $\delta(t)$ is the unit impulse function. Show that

$$y(t) = \cos(\sqrt{5}t) + \frac{u_2}{2\sqrt{5}} \sin(\sqrt{5}(t - 2)),$$

if $y(0) = 1, \quad y'(0) = 0.$

Exercise 4.8. Assume that the deflection of a beam is governed by the following equation:

$$EIy^{(4)}(x) = \frac{W_0}{c} [c - x + (x - c)u_c], \quad 0 < x < 2c.$$

Solve this equation under the boundary conditions

$$y(0) = y'(0) = y''(2c) = y'''(2c) = 0.$$

Exercise 4.9. The differential equation

$$Li''(t) + \frac{1}{C}i = v'(t)$$

arises from Kirchhoff's second law for an electrical circuit. Solve this equation under the initial conditions

$$i(0) = i'(0) = 0,$$

where $v(t) = \begin{cases} -4t, & 0 \le t \le 3 \\ -12, & t > 3 \end{cases}.$

Exercise 4.10. Suppose that the driving voltage for an RC-circuit is $\sin(t)$ for $t \geq 0$. Find the current of the circuit if the initial current is $1/R$ and the equation governing the circuit is $Ri'(t) + \dfrac{1}{C} \displaystyle\int_0^t i(x)dx = \sin(t)$.

Exercise 4.11. A simple satellite-tracking system is modeled by the differential equation

$$I\theta''(t) = -k[\theta(t) - \Psi(t)].$$

In this expression $\Psi(t)$ is an observed input angle, $\theta(t)$ is the system response output, I is an appropriate moment of inertia, and k denotes a positive constant that relates the system's induced compensating torque to the instantaneous pointing $\theta(t) - \Psi(t)$. If $\Psi(t)$ is assumed to be known, determine the solution of the initial value problem with

$$\theta(0) = \theta_0, \quad \theta'(0) = \theta_0'$$

by the Laplace transform method and show that it can be expressed in convolution form as

$$\theta(t) = \theta_0 \cos(\omega t) + \left(\frac{\theta_0'}{\omega}\right) \sin(\omega t) + \omega \int_0^t \Psi(u) \sin(t - u)du,$$

where $\omega \equiv \sqrt{k/I}$.

Chapter 5
Systems of Linear Differential Equations

5.1 Introduction

Systems of linear ODEs that arise frequently as mathematical models of mechanical or electrical systems contain several dependent variables $x_1, \ldots, x_n$, where each of them is a function of the single independent variable t. One of the mathematical reasons to study such a system of linear ODEs is the connection between the system of differential equations and the linear nth-order ODE:

$$y^{(n)} + p_1(t)y^{(n-1)} + p_2(t)y^{(n-2)} + \cdots + p_{n-1}(t)y' + p_n(t)y = f(t).$$

To see this, let

$$x_1 \equiv y, \quad x_2 \equiv y', \ldots, x_n \equiv y^{(n-1)}.$$

Then, the given scalar equation is equivalent to the system of ODEs:

$$x_1' = x_2,$$

$$x_2' = x_3,$$

$$\vdots$$

$$x_{n-1}' = x_n,$$

$$x_n' = f(t) - p_n(t)x_1 - p_{n-1}(t)x_2 - \cdots - p_1(t)x_n.$$

Obviously, the actual information of the nth-order ODE is contained in the nth equation of the system. The first $n - 1$ equations are trivially satisfied.

M. Hermann and M. Saravi, *A First Course in Ordinary Differential Equations: Analytical and Numerical Methods*, DOI 10.1007/978-81-322-1835-7_5, © Springer India 2014

The more general form of a system of ODEs is

$$
\begin{aligned}
x_1' &= f_1(t, x_1, x_2, \ldots, x_n), \\
x_2' &= f_2(t, x_1, x_2, \ldots, x_n), \\
&\ \vdots \\
x_n' &= f_n(t, x_1, x_2, \ldots, x_n).
\end{aligned}
\tag{5.1}
$$

The ODEs (5.1) together with the initial conditions

$$
x_1(t_0) = x_{01}, \quad x_2(t_0) = x_{02}, \quad \ldots \quad , x_n(t_0) = x_{0n}
\tag{5.2}
$$

form the associated initial value problem.

The system of differential equations (5.1) may be linear or nonlinear. In this chapter we consider only linear systems which are of the form

$$
\begin{aligned}
x_1' &= a_{11}x_1 + a_{12}x_2 + \cdots + a_{1n}x_n + f_1(t), \\
x_2' &= a_{21}x_1 + a_{22}x_2 + \cdots + a_{2n}x_n + f_2(t), \\
&\ \vdots \\
x_n' &= a_{n1}x_1 + a_{n2}x_2 + \cdots + a_{nn}x_n + f_n(t),
\end{aligned}
\tag{5.3}
$$

where a_{ij} are functions of t. The system (5.3) can be written in vector notation as

$$
x'(t) = A(t)x(t) + f(t),
\tag{5.4}
$$

where

$$
x'(t) \equiv (x_1', x_2', \ldots, x_n')^T,
$$

$$
x(t) \equiv (x_1, x_2, \ldots, x_n)^T,
$$

$$
f(t) \equiv (f_1, f_2, \ldots, f_n)^T,
$$

and $A(t)$ is the corresponding coefficient matrix.

For the system (5.4), the existence and uniqueness of solutions can be stated analogous to the theorems in Chaps. 2 and 3.

Theorem 5.1. *If the functions a_{ij} are continuous on the interval $a < t < b$ containing the point t_0, then there exists a unique solution $x(t)$ of the system (5.3), which satisfies the initial conditions (5.2).*

In this chapter we will study the system (5.4). First, we consider the method of solution by elimination utilizing the D operator and Laplace transforms. With the use of Laplace transforms, a system of ODEs is transferred into a system of algebraic equations that can be solved by an elimination method. Our second method is connected to theory of linear algebra, because we need to solve a corresponding algebraic eigenvalue and eigenvector problem.

5.2 Elimination Methods

5.2.1 Elimination Using the D Operator

Consider the system of equations

$$p_{11}x_1 + p_{12}x_2 + \cdots + p_{1n}x_n = f_1(t),$$

$$p_{21}x_1 + p_{22}x_2 + \cdots + p_{2n}x_n = f_2(t),$$

$$\vdots$$

$$p_{n1}x_1 + p_{n2}x_2 + \cdots + p_{nn}x_n = f_n(t),$$

(5.5)

where p_{ij} are polynomials in the linear operator $D = d/dt$ and obey the rules of algebra. We can solve this system in a similar way as a system of simultaneous algebraic equation is solved by the elimination method.

Example 5.2. Solve the system of equations given by

$$2\frac{dx}{dt} - 3\frac{dy}{dt} = 0,$$

$$\frac{dx}{dt} - 2\frac{dy}{dt} = 2t.$$

Solution. We write

$$2Dx - 3Dy = 0, \quad Dx - 2Dy = 2t.$$

We multiply the first equation by -2 and the second one by 3 and add them. We get $Dx = -6t$. By integration, it is easy to show that $x = -3t^2 + c_1$. To find y, this expression for x is substituted into one of these equations. It is not difficult to obtain $y(t) = -2t^2 + c_2$.

$\square$

Example 5.3. Solve

$$x' = x + 2y,$$

$$y' = 2x + 4y.$$

Solution. Let us formulate this system with the D operator:

$$Dx = x + 2y, \quad Dy = 2x + 4y,$$

or

$$(D - 1)x - 2y = 0, \quad 2x + (D - 4)y = 0.$$

Multiplying the first equation by $(D - 4)$ and the second one by 2 and adding them lead to $D(D - 5)x = 0$. It is easy to show that the general solution of this equation will be $x(t) = c_1 + c_2 e^t$. By substituting x into one of these two equations, we obtain $y(t) = c_3 + c_4 e^t$. This is not a coincidence that the solutions are the same. Since the determinant of the coefficient matrix is zero, we obtain equal solutions. In the next examples, we try to avoid this situation.

□

Example 5.4. Solve

$$(D - 2)x + 2Dy = 2(1 - 2e^{2t}), \quad (2D - 3)x + (3D - 1)y = 0.$$

Solution. To eliminate y, we multiply the first equation by $(3D-1)$ and the second equation by $-2D$ and add them. We obtain

$$(3D - 1)(D - 2)x - 2D(2D - 3)x = 2(3D - 1)(1 - 2e^{2t}).$$

Grouping powers of D leads to

$$(D^2 + D - 2)x = 2(1 + 10e^{2t}).$$

It is easy to show that $x_c(t) \equiv c_1 e^t + c_2 e^{-2t}$ and $x_p(t) \equiv 5e^{2t} - 1$. Hence

$$x(t) = c_1 e^t + c_2 e^{-2t} + 5e^{2t} - 1.$$

Substituting x into one of the two equations yields

$$y(t) = \frac{c_1}{2} e^t - c_2 e^{-2t} - e^{2t} + 3.$$

□

Remark 5.5. One can also use Cramer's rule to solve the system of differential equations if the determinant of the coefficient matrix is not zero and the dimension of the problem is small. □

5.2.2 Elimination Using Laplace Transforms

The Laplace transforms can also be used to solve systems of ODEs with constant coefficients. The procedure is essentially the same as that described in the previous section.

Example 5.6. Suppose a mechanical system contains two masses on three springs and is governed by the differential equations

$$x_1'' = k(x_2 - 2x_1),$$

$$x_2'' = k(x_1 - 2x_2),$$

where k is the spring modulus of each of the three springs, and x_1 and x_2 are the displacements of the masses from their position of static equilibrium. The masses of the springs and the damping have been neglected. Solve this system with the initial conditions

$$x_1(0) = x_2(0) = 0 \quad \text{and} \quad x_1'(0) = x_2'(0) = 1.$$

Solution. We use the $\mathcal{L}$ operator instead of the D operator. Note that both operators are linear. We write

$$\mathcal{L}\{x_1''\} = k\mathcal{L}(x_2 - 2x_1),$$

and we obtain

$$s^2\mathcal{L}\{x_1\} - sx_1(0) - x_1'(0) = k(\mathcal{L}\{x_2\} - 2\mathcal{L}\{x_1\}).$$

For the second ODE, we have

$$\mathcal{L}\{x_2''\} = k\mathcal{L}(x_1 - 2x_2),$$

and we obtain

$$s^2\mathcal{L}\{x_2\} - sx_2(0) - x_2'(0) = k(\mathcal{L}\{x_1\} - 2\mathcal{L}\{x_2\}).$$

Hence

$$(s^2 + 2)\mathcal{L}\{x_1\} - k\mathcal{L}\{x_2\} = 1,$$

$$-k\mathcal{L}\{x_1\} + (s^2 + 2)\mathcal{L}\{x_2\} = 1.$$

Eliminating $\mathcal{L}\{x_2\}$ from these two equations yields

$$\mathcal{L}\{x_1\} = \frac{1}{s^2 + 2 - k}.$$

Thus,

$$x_1(t) = \mathcal{L}^{-1}\left\{\frac{1}{s^2 + 2 - k}\right\} = \frac{1}{\sqrt{2-k}}\sin(\sqrt{2-k}\,t).$$

Now we find

$$x_2(t) = \frac{1}{\sqrt{2-k}}\sin(\sqrt{2-k}\,t).$$

□

Example 5.7. Solve the initial value problem

$$x'' - 4x - y' = t,$$

$$10x' + y'' - y = -1,$$

with

$$x(0) = y(0) = x'(0) = 0.$$

Solution. The Laplace transforms of each equation lead to

$$(s^2 - 4)\mathcal{L}\{x\} - s\mathcal{L}\{y\} = \frac{1}{s^2},$$

$$10s\mathcal{L}\{x\} + (s^2 - 1)\mathcal{L}\{y\} = -\frac{1}{s}.$$

Elimination of $\mathcal{L}\{y\}$ yields

$$\mathcal{L}\{x\} = -\frac{1}{s^2(s^2 + 4)(s^2 + 1)} = -\frac{1/4}{s^2} + \frac{1/3}{(s^2 + 4)} - \frac{1/12}{s^2 + 1}.$$

Therefore,

$$x(t) = -\mathcal{L}^{-1}\left\{\frac{1/4}{s^2}\right\} + \mathcal{L}^{-1}\left\{\frac{1/3}{s^2 + 4}\right\} - \mathcal{L}^{-1}\left\{\frac{1/12}{s^2 + 1}\right\}$$

$$= -\frac{t}{4} + \frac{1}{6}\sin(2t) - \frac{1}{12}\sin(t).$$

Substituting $\mathcal{L}\{x\}$ into the first equation and a similar calculation leads to

$$y(t) = \frac{2}{3}\cos(2t) - \frac{5}{12}\cos(t) - \frac{1}{4}.$$

□

Remark 5.8. It is not necessary that initial conditions are known. We can impose arbitrary initial conditions and find the solutions in terms of these arbitrary initial values.

□

5.3 Method of Vector-Matrix Notation

The elimination method discussed in the preceding sections is useful when the number of equation is small. But, if a large system is to be considered, then it is reasonable to apply some results of the theory of matrices. We will assume, however, that the readers are familiar with eigenvalues and eigenvectors of matrices and how to evaluate them.

In Sect. 5.1 we pointed out that any scalar nth-order ODE can be written as a system of n first-order differential equations. In this section, we consider a system of n first-order linear ODEs with constant coefficients. The general form of such a system is given by

$$x'(t) = Ax(t) + f(t), \quad t \in [a, b], \tag{5.6}$$

where $x \equiv (x_1, x_2, \ldots, x_n)^T$, $f \equiv (f_1, f_2, \ldots, f_n)^T$, and A is a constant $n \times n$ coefficient matrix. The system is homogeneous if $f(t) = 0$. To determine the solution of (5.6), we first turn to the associated homogeneous system. That is,

$$x'(t) = Ax(t). \tag{5.7}$$

Theorem 5.9. *If the vector functions $x_1(t), x_2(t), \ldots, x_n(t)$ are linearly independent solutions (a fundamental set) of the equation (5.7) on the interval $a < t < b$, then*

$$x(t) = c_1 x_1(t) + c_2 x_2(t) + \cdots + c_n x_n(t) \tag{5.8}$$

is also a solution of (5.7) for any constants $c_1, c_2, \ldots, c_n$.

Proof. See, e.g., [1]. ∎

We call $x(t)$ the *general solution* of (5.7). Let $x_1(t), x_2(t), \ldots, x_n(t)$ be a fundamental set of solutions of (5.7). Then the matrix

$$X(t) \equiv (x_1(t) \mid x_2(t) \mid \cdots \mid x_n(t)) \in \mathbb{R}^{n \times n} \tag{5.9}$$

is said to be a *fundamental matrix*. This matrix is nonsingular since its columns are linearly independent vectors. It is not difficult to show that the general solution of (5.7) is given by

$$x(t) = X(t)\,\mathbf{c}, \quad \mathbf{c} \in \mathbb{R}^n. \tag{5.10}$$

It is obvious that by analogy with the treatment of second-order linear ODEs in Chap. 3, we will seek the solutions of (5.7) in the form

$$x(t) = e^{mt}\,\boldsymbol{\varphi}, \tag{5.11}$$

where $\boldsymbol{\varphi} \in \mathbb{R}^n$ is a constant vector and m is a real number. Substituting the ansatz (5.11) into (5.7), we obtain

$$m\,e^{mt}\,\boldsymbol{\varphi} = e^{mt}\,A\,\boldsymbol{\varphi}, \text{ i.e. } e^{mt}(A - mI)\boldsymbol{\varphi} = \mathbf{0},$$

where I is the $n \times n$ identity matrix. Since $e^{mt} \neq 0$, we get

$$(A - mI)\,\boldsymbol{\varphi} = 0. \tag{5.12}$$

A solution $x(t) \not\equiv \mathbf{0}$ of the differential equation (5.7) is called a nontrivial solution. In order to determine a nontrivial solution of system (5.7), we must solve the system of algebraic equations (5.12). This latter problem is precisely the one to determine the eigenvalues and eigenvectors of A. Thus, the vector $x(t)$ given by (5.11) is a solution of Eq. (5.7) provided m is an eigenvalue and φ is an associated eigenvector of A. But we know from elementary linear algebra that Eq. (5.7) has a nontrivial solution if and only if m is chosen so that $\det(A - mI) = 0$. This leads to an nth-order polynomial equation (*auxiliary equation*) with n roots. After the determination of an eigenvalue using the auxiliary equation, the corresponding eigenvector may be evaluated by solving a homogeneous system of n linear algebraic equations in n unknowns.

The eigenvalues can be real or complex numbers. Since the auxiliary equation can have simple or multiple roots, we also must differ between simple and multiple eigenvalues. In each of these cases, the general solution can be constructed similar to the determination of the general solution of a linear nth-order ODE. We continue this section with some examples.

5.3.1   Simple Real Eigenvalues

Assume $m_1, m_2, \ldots, m_n \in \mathbb{R}$ are distinct roots of the polynomial

$$\det(A - mI) = 0,$$

i.e., the eigenvalues of the matrix A. Let $\varphi_1, \varphi_2, \ldots, \varphi_n$ be n linearly independent eigenvectors associated with these eigenvalues. Then

$$x_1(t) = e^{m_1 t}\, \varphi_1, \quad x_2(t) = e^{m_2 t}\, \varphi_2, \ldots, x_n(t) = e^{m_n t}\, \varphi_n$$

are n linearly independent solutions of (5.7). Therefore, the general solution is

$$x(t) = c_1 e^{m_1 t} \varphi_1 + c_2 e^{m_2 t} \varphi_2 + \cdots + c_n e^{m_n t} \varphi_n. \tag{5.13}$$

Note that if A is real and symmetric, then the eigenvectors as well as the eigenvalues are all real. However, if A is symmetric but is not real, then in general the eigenvectors have nonzero imaginary parts, and the solutions are complex values. We also note that if A is symmetric, then the eigenvectors are orthogonal to each other, i.e., $\varphi_i^T \varphi_j = 0, i \neq j$.

Example 5.10. Find the general solution of

$$x'(t) = \begin{pmatrix} 1 & 2 \\ 2 & 1 \end{pmatrix} x(t).$$

Solution. Since the coefficient matrix is real and symmetric, the eigenvalues and the corresponding eigenvectors are real. We have

$$\det(A - mI) = \det \begin{pmatrix} 1-m & 2 \\ 2 & 1-m \end{pmatrix} = 0, \quad \text{i.e. } m^2 - 2m - 3 = 0.$$

Thus, $m_1 = 3, m_2 = -1$. For finding the corresponding eigenvectors, we have to consider the eigenvalue problems

$$A\varphi_i = m_i \varphi_i.$$

Hence

$$A\varphi_1 = m_1\varphi_1 \ : \ \begin{pmatrix} 1 & 2 \\ 2 & 1 \end{pmatrix}\begin{pmatrix} \varphi_{1,1} \\ \varphi_{1,2} \end{pmatrix} = 3\begin{pmatrix} \varphi_{1,1} \\ \varphi_{1,2} \end{pmatrix} \ \rightarrow \ \varphi_1 = \begin{pmatrix} 1 \\ 1 \end{pmatrix},$$

$$A\varphi_2 = m_2\varphi_2 \ : \ \begin{pmatrix} 1 & 2 \\ 2 & 1 \end{pmatrix}\begin{pmatrix} \varphi_{2,1} \\ \varphi_{2,2} \end{pmatrix} = -\begin{pmatrix} \varphi_{2,1} \\ \varphi_{2,2} \end{pmatrix} \ \rightarrow \ \varphi_2 = \begin{pmatrix} -1 \\ 1 \end{pmatrix}.$$

Then,

$$x_1(t) = e^{m_1 t}\varphi_1 = e^{3t}\begin{pmatrix} 1 \\ 1 \end{pmatrix}, \quad x_2(t) = e^{m_2 t}\varphi_2 = e^{-t}\begin{pmatrix} -1 \\ 1 \end{pmatrix},$$

and the general solution is

$$x(t) = c_1\,x_1(t) + c_2\,x_2(t) = c_1\,e^{3t} \begin{pmatrix} 1 \\ 1 \end{pmatrix} + c_2\,e^{-t} \begin{pmatrix} -1 \\ 1 \end{pmatrix}.$$

Note that $x_1^T x_2 = 0$.

$\square$

Example 5.11. Solve the system of ODEs $x'(t) = Ax(t)$, where

$$A = \begin{pmatrix} 1 & 2 & -1 \\ 0 & 2 & 1 \\ 0 & 0 & 3 \end{pmatrix}.$$

Solution. From $\det(A - mI) = 0$ we obtain

$$\det \begin{pmatrix} 1-m & 2 & -1 \\ 0 & 2-m & 1 \\ 0 & 0 & 3-m \end{pmatrix} = (1-m)(2-m)(3-m) = 0.$$

It is obvious that $m_1 = 1$, $m_2 = 2$ and $m_3 = 3$. From $A\varphi_1 = m_1\varphi_1$, we have

$$\begin{pmatrix} 1 & 2 & -1 \\ 0 & 2 & 1 \\ 0 & 0 & 3 \end{pmatrix} \begin{pmatrix} \varphi_{1,1} \\ \varphi_{1,2} \\ \varphi_{1,3} \end{pmatrix} = 1 \begin{pmatrix} \varphi_{1,1} \\ \varphi_{1,2} \\ \varphi_{1,3} \end{pmatrix}.$$

Thus, the first eigenvector is

$$\varphi_1 = (1,0,0)^T.$$

In a similar manner, from $A\varphi_2 = m_2\varphi_2$, we get

$$\begin{pmatrix} 1 & 2 & -1 \\ 0 & 2 & 1 \\ 0 & 0 & 3 \end{pmatrix} \begin{pmatrix} \varphi_{2,1} \\ \varphi_{2,2} \\ \varphi_{2,3} \end{pmatrix} = 2 \begin{pmatrix} \varphi_{2,1} \\ \varphi_{2,2} \\ \varphi_{2,3} \end{pmatrix}.$$

The second eigenvector is

$$\varphi_2 = (2,1,0)^T.$$

Finally, from $A\varphi_3 = m_3\varphi_3$ we obtain

$$\begin{pmatrix} 1 & 2 & -1 \\ 0 & 2 & 1 \\ 0 & 0 & 3 \end{pmatrix} \begin{pmatrix} \varphi_{3,1} \\ \varphi_{3,2} \\ \varphi_{3,3} \end{pmatrix} = 3 \begin{pmatrix} \varphi_{3,1} \\ \varphi_{3,2} \\ \varphi_{3,3} \end{pmatrix}.$$

Obviously, the third eigenvector is

$$\boldsymbol{\varphi}_3 = (1, 2, 2)^T.$$

Hence the general solution is

$$x(t) = c_1 \, x_1(t) + c_2 \, x_2(t) + c_3 \, x_3(t)$$

$$= c_1 \, e^t \begin{pmatrix} 1 \\ 0 \\ 0 \end{pmatrix} + c_2 \, e^{2t} \begin{pmatrix} 2 \\ 1 \\ 0 \end{pmatrix} + c_3 \, e^{3t} \begin{pmatrix} 1 \\ 2 \\ 2 \end{pmatrix}.$$

□

5.3.2   Multiple Real Eigenvalues

Suppose the auxiliary equation has a multiple eigenvalue. For example, we have a 2×2 matrix with $m_1 = m_2 = m$. Clearly, $x_1(t) = e^{mt} \boldsymbol{\varphi}_1$. Our experience with multiple roots on solving second-order linear ODEs with constant coefficients helps us to seek an additional solution in the form $x_2(t) = t e^{mt} \boldsymbol{\varphi}_2$. But a simple substitution leads to $\boldsymbol{\varphi}_2 = \mathbf{0}$. This cannot be acceptable because we are looking for a nonzero solution. Thus, we assume

$$x_2(t) = e^{mt} \boldsymbol{\varphi}(t) = e^{mt} \begin{pmatrix} \varphi_1(t) \\ \varphi_2(t) \end{pmatrix}.$$

If we enforce this representation, we obtain

$$x_2(t) = e^{mt} (t \boldsymbol{\varphi}_1 + \boldsymbol{\varphi}_2).$$

To have more understanding, we consider two examples:

Example 5.12. Solve the initial value problem given by

$$x'(t) = \begin{pmatrix} 3 & -1 \\ 1 & 5 \end{pmatrix} x(t), \quad x(0) = \begin{pmatrix} 1 \\ 1 \end{pmatrix}.$$

Solution. It is

$$\det \begin{pmatrix} 3 - m & -1 \\ 1 & 5 - m \end{pmatrix} = m^2 - 8m + 16 = (m - 4)^2 = 0.$$

The roots of this auxiliary equation are $m_1 = m_2 = 4$. Thus, we have a multiple (double) eigenvalue. From the eigenvalue problem $A\varphi_1 = 4\varphi_1$, we obtain

$$\varphi_1 = \begin{pmatrix} 1 \\ -1 \end{pmatrix}, \quad \text{i.e.} \quad x_1(t) = e^{4t} \begin{pmatrix} 1 \\ -1 \end{pmatrix}.$$

For the second solution, we use the ansatz

$$x_2(t) = e^{4t} \left(t \begin{pmatrix} 1 \\ -1 \end{pmatrix} + \varphi_2 \right).$$

A simple substitution leads to

$$\varphi_2 = \begin{pmatrix} -1 \\ 0 \end{pmatrix}, \quad \text{i.e.} \quad x_2(t) = e^{4t} \left(t \begin{pmatrix} 1 \\ -1 \end{pmatrix} + \begin{pmatrix} -1 \\ 0 \end{pmatrix} \right).$$

Therefore, the general solution is

$$x(t) = c_1 e^{4t} \begin{pmatrix} 1 \\ -1 \end{pmatrix} + c_2 e^{4t} \left(t \begin{pmatrix} 1 \\ -1 \end{pmatrix} + \begin{pmatrix} -1 \\ 0 \end{pmatrix} \right).$$

If we impose the given initial condition, one can easily show that $c_1 = -1$ and $c_2 = -2$.

$\square$

Example 5.13. Find the general solution of

$$x'(t) = \begin{pmatrix} 1 & 1 \\ -1 & 3 \end{pmatrix} x(t).$$

Solution. Since the coefficient matrix is real and symmetric, the eigenvalues and associated eigenvectors are real.

We have

$$\det \begin{pmatrix} 1-m & 1 \\ -1 & 3-m \end{pmatrix} = (m-2)^2 = 0, \quad \text{i.e.} \quad m_1 = m_2 = 2.$$

From $A\varphi_1 = 2\varphi_1$, we obtain

$$\varphi_1 = \begin{pmatrix} 1 \\ 1 \end{pmatrix}, \quad \text{i.e.} \quad x_1(t) = e^{2t} \begin{pmatrix} 1 \\ 1 \end{pmatrix}.$$

For the second solution, we set

$$x_2(t) = e^{2t} \left(t \begin{pmatrix} 1 \\ 1 \end{pmatrix} + \varphi_2 \right).$$

If we substitute this ansatz into the given system, we obtain

$$\begin{pmatrix} 1 + 2t + 2\varphi_{2,1} \\ 1 + 2t + 2\varphi_{2,2} \end{pmatrix} = \begin{pmatrix} 2t + \varphi_{2,1} + \varphi_{2,2} \\ 2t - \varphi_{2,1} + 3\varphi_{2,2} \end{pmatrix}, \quad \text{i.e.} \quad \varphi_{2,2} - \varphi_{2,1} = 1.$$

Let us choose $\varphi_{2,1} = 2$ and $\varphi_{2,1} = 1$. Then $\boldsymbol{\varphi}_2 = \begin{pmatrix} 1 \\ 2 \end{pmatrix}$ and the general solution is

$$\boldsymbol{x}(t) = c_1 e^{2t} \begin{pmatrix} 1 \\ 1 \end{pmatrix} + c_2 e^{2t} \left(t \begin{pmatrix} 1 \\ 1 \end{pmatrix} + \begin{pmatrix} 1 \\ 2 \end{pmatrix} \right).$$

$\square$

5.3.3   Simple Complex Eigenvalues

In this case the general solution is the same as given in (5.12), but m is a complex number. Since any complex roots must occur in conjugate pairs, then by analogy with the treatment of second-order linear ODEs, we might use Euler's formula to represent the general solution with trigonometric functions.

Example 5.14. Solve the system of ODEs $\boldsymbol{x}'(t) = A\boldsymbol{x}(t)$, where

$$A = \begin{pmatrix} 1 & -2 \\ 2 & 1 \end{pmatrix}.$$

Solution. We have

$$\det \begin{pmatrix} 1 - m & -2 \\ 2 & 1 - m \end{pmatrix} = (1 - m)^2 + 4 = 0, \quad \text{i.e.} \quad m = 1 \pm 2i.$$

Suppose $m_1 = 1 + 2i$. The corresponding eigenvalue problem is

$$A\boldsymbol{\varphi}_1 = (1 + 2i)\boldsymbol{\varphi}_1, \quad \text{i.e.} \quad \boldsymbol{\varphi}_1 = \begin{pmatrix} 1 \\ -i \end{pmatrix}.$$

Similarly from $A\boldsymbol{\varphi}_2 = (1 - 2i)\boldsymbol{\varphi}_2$, we obtain $\boldsymbol{\varphi}_2 = \begin{pmatrix} 1 \\ i \end{pmatrix}$. Hence

$$\boldsymbol{x}(t) = c_1 e^{(1+2i)t} \begin{pmatrix} 1 \\ -i \end{pmatrix} + c_2 e^{(1-2i)t} \begin{pmatrix} 1 \\ i \end{pmatrix}.$$

By using Euler's formula

$$e^{\pm i\theta} = \cos(\theta) \pm i\,\sin(\theta),$$

we may write

$$x(t) = e^t \left[c_1 e^{2it} \begin{pmatrix} 1 \\ -i \end{pmatrix} + c_2 e^{-2it} \begin{pmatrix} 1 \\ i \end{pmatrix} \right]$$

$$= e^t \left[c_1 (\cos(2t) + i\,\sin(2t)) \begin{pmatrix} 1 \\ -i \end{pmatrix} + c_2(\cos(2t) - i\,\sin(2t)) \begin{pmatrix} 1 \\ i \end{pmatrix} \right]$$

$$= e^t \left\{ A \left[\cos(2t) \begin{pmatrix} 1 \\ 0 \end{pmatrix} + \sin(2t) \begin{pmatrix} 0 \\ 1 \end{pmatrix} \right] \right.$$

$$\left. + B \left[\cos(2t) \begin{pmatrix} 0 \\ -1 \end{pmatrix} + \sin(2t) \begin{pmatrix} 1 \\ 0 \end{pmatrix} \right] \right\},$$

where $A \equiv c_1 + c_2$ and $B \equiv i(c_1 - c_2)$.

$\square$

5.4 Solution of an Inhomogeneous System of ODEs

Let us now consider systems of the form

$$x'(t) = A\,x(t) + f(t), \quad t \in [a, b], \tag{5.14}$$

where

$$x(t) = (x_1, x_2, \ldots, x_n)^T, \quad f(t) = (f_1, f_2, \ldots, f_n)^T,$$

and $A \in \mathbb{R}^{n \times n}$ is a constant coefficient matrix. We assume $f(t) \not\equiv 0$. The general solution of the associated homogeneous system of (5.14) can be represented with the help of a fundamental matrix X. Thus, from (5.10), we have

$$x_{\text{hom}}(t) = X(t)\,c. \tag{5.15}$$

Now, we use (5.15) to find a particular solution of (5.14) by the method of variation of parameters. Hence, we assume

$$x_p(t) = X(t)c(t) \tag{5.16}$$

is a solution of (5.14), where $c(t) = (c_1(t), \ldots, c_n(t))^T$. Differentiating (5.16) leads to

$$x'_p(t) = X'(t)c(t) + X(t)c'(t). \tag{5.17}$$

But $X'(t) = AX(t)$. Hence, from Eq. (5.17) we obtain

$$x'_p(t) = AX(t)c(t) + X(t)c'(t). \tag{5.18}$$

From (5.14), we also have

$$x'_p(t) = AX(t)c(t) + f(t). \tag{5.19}$$

Comparing equations (5.18) and (5.19) implies that

$$X(t)c'(t) = f(t), \text{ i.e. } c'(t) = X^{-1}(t)f(t). \tag{5.20}$$

Here we have used the fact that $X(t)$ is a fundamental matrix, and therefore the inverse $X^{-1}(t)$ exists. After the integration we get

$$c(t) = \int_{t_0}^t X^{-1}(\tau)f(\tau)d\tau + k,$$

where $k \in \mathbb{R}^n$ is the vector of integration constants. Substituting $c(t)$ into (5.16), the following representation for the particular solution is obtained:

$$x_p(t) = \int_{t_0}^t X(t)X^{-1}(\tau)f(\tau)d\tau + X(t)k. \tag{5.21}$$

Since $x_p(t)$ is a particular solution, we can assume that it satisfies the special initial condition $x_p(t_0) = 0$. If the fundamental matrix is chosen such that $X(t_0) = I$, $I \in \mathbb{R}^{n \times n}$ identity matrix, we have immediately $k = 0$. Now, using the superposition principle, the general solution of the inhomogeneous system (5.14) is

$$x(t) = x_{\text{hom}}(t) + x_p(t) = X(t)c + \int_{t_0}^t X(t)X^{-1}(\tau)f(\tau)d\tau. \tag{5.22}$$

Finally, if an initial condition $x(t_0) = x_0$ is added to the ODE (5.14), we have $c = x_0$.

Example 5.15. Find a particular solution of

$$x'(t) = \begin{pmatrix} 1 & 2 \\ 2 & 1 \end{pmatrix} x(t) + \begin{pmatrix} e^{2t} \\ e^t \end{pmatrix}.$$

Solution. Using the result of Example 5.10, we have the following solution of the corresponding homogeneous problem:

$$x_{\text{hom}}(t) = c_1 e^{3t} \begin{pmatrix} 1 \\ 1 \end{pmatrix} + c_2 e^{-t} \begin{pmatrix} -1 \\ 1 \end{pmatrix}.$$

Therefore, the fundamental matrix is

$$X(t) = \begin{pmatrix} e^{3t} & -e^{-t} \\ e^{3t} & e^{-t} \end{pmatrix}.$$

The inverse of this matrix is

$$X^{-1}(t) = \frac{1}{2} \begin{pmatrix} e^{-3t} & e^{-3t} \\ -e^{t} & e^{t} \end{pmatrix}.$$

Using formula (5.20), we obtain

$$c'(t) = \frac{1}{2} \begin{pmatrix} e^{-3t} & e^{-3t} \\ -e^{t} & e^{t} \end{pmatrix} \begin{pmatrix} e^{2t} \\ e^{t} \end{pmatrix} = \frac{1}{2} \begin{pmatrix} e^{-t} + e^{-2t} \\ -e^{3t} + e^{2t} \end{pmatrix}.$$

Thus,

$$c(t) = \frac{1}{2} \begin{pmatrix} -e^{-t} - \dfrac{1}{2} e^{-2t} \\ -\dfrac{1}{3} e^{3t} + \dfrac{1}{2} e^{2t} \end{pmatrix}.$$

Substituting $c(t)$ into (5.16), we get

$$x_p(t) = \frac{1}{2} \begin{pmatrix} e^{3t} & -e^{-t} \\ e^{3t} & e^{-t} \end{pmatrix} \begin{pmatrix} -e^{-t} - \dfrac{1}{2} e^{-2t} \\ -\dfrac{1}{3} e^{3t} + \dfrac{1}{2} e^{2t} \end{pmatrix}$$

$$= \begin{pmatrix} -\dfrac{1}{3} e^{2t} - \dfrac{1}{2} e^{t} \\ -\dfrac{2}{3} e^{2t} \end{pmatrix}.$$

$\square$

Remark 5.16. In the case $n = 2$, the ansatz (5.16) for the particular solution can be written in the form

$$x_p(t) = c_1(t)x_1(t) + c_2(t)x_2(t),$$

where $x_1(t)$ and $x_2(t)$ are the columns of the fundamental matrix X (see formula (5.9)). $\square$

Example 5.17. Find the general solution of

$$x'(t) = \begin{pmatrix} 4 & 3 \\ -4 & -4 \end{pmatrix} x(t) + \begin{pmatrix} 2 \\ e^{3t} \end{pmatrix}.$$

Solution. First we solve the associated homogeneous system, that is,

$$x'(t) = \begin{pmatrix} 4 & 3 \\ -4 & -4 \end{pmatrix} x(t).$$

It is not difficult to show that the solution is

$$x_{\text{hom}}(t) = c_1 e^{2t} \begin{pmatrix} 3 \\ -2 \end{pmatrix} + c_2 e^{-2t} \begin{pmatrix} 1 \\ -2 \end{pmatrix}.$$

Now we use the following ansatz for the particular solution (see Remark 5.16):

$$x_p(t) = c_1(t) e^{2t} \begin{pmatrix} 3 \\ -2 \end{pmatrix} + c_2(t) e^{-2t} \begin{pmatrix} 1 \\ -2 \end{pmatrix}.$$

It is

$$x'_p(t) = c'_1(t) e^{2t} \begin{pmatrix} 3 \\ -2 \end{pmatrix} + 2c_1(t) e^{2t} \begin{pmatrix} 3 \\ -2 \end{pmatrix}$$

$$+ c'_2(t) e^{-2t} \begin{pmatrix} 1 \\ -2 \end{pmatrix} - 2c_2(t) e^{-2t} \begin{pmatrix} 1 \\ -2 \end{pmatrix}.$$

Substituting $x_p(t)$ and $x'_p(t)$ into the given system, we obtain

$$c'_1(t) e^{2t} \begin{pmatrix} 3 \\ -2 \end{pmatrix} + c'_2(t) e^{-2t} \begin{pmatrix} 1 \\ -2 \end{pmatrix} = \begin{pmatrix} 2 \\ e^{3t} \end{pmatrix}.$$

The component-wise representation is

$$3c'_1(t) e^{2t} + c'_2(t) e^{-2t} = 2,$$

$$-2c'_1(t) e^{2t} - 2c'_2(t) e^{-2t} = e^{3t}.$$

(5.23)

Eliminating $c'_2(t)$, we obtain

$$c'_1(t) = \frac{1}{4} \left(4e^{-2t} + e^t \right), \text{ i.e. } c_1(t) = \frac{1}{4} \left(-2e^{-2t} + e^t \right).$$

Substituting $c_1'(t)$ into the first equation of (5.23) gives

$$c_2'(t) = -e^{2t} - \frac{3}{4}e^{5t}, \quad \text{i.e.} \quad c_2(t) = -\frac{1}{2}e^{2t} - \frac{3}{20}e^{5t}.$$

Thus, the particular solution is

$$x_p(t) = \frac{1}{4}\left(-2 + e^{3t}\right)\begin{pmatrix} 3 \\ -2 \end{pmatrix} - \frac{1}{2}\left(1 + \frac{3}{10}e^{3t}\right)\begin{pmatrix} 1 \\ -2 \end{pmatrix}.$$

Using the superposition principle, the general solution of the given ODE can be represented in the form

$$x(t) = x_{\text{hom}} + x_p(t)$$

$$= c_1 e^{2t}\begin{pmatrix} 3 \\ -2 \end{pmatrix} + c_2 e^{-2t}\begin{pmatrix} 1 \\ -2 \end{pmatrix}$$

$$+ \frac{1}{4}\left(-2 + e^{3t}\right)\begin{pmatrix} 3 \\ -2 \end{pmatrix} - \frac{1}{2}\left(1 + \frac{3}{10}e^{3t}\right)\begin{pmatrix} 1 \\ -2 \end{pmatrix}.$$

□

5.5   Method of Undetermined Coefficients

For obtaining the particular solution by the method of variation of parameters, we need to know the general solution of the associated homogeneous system. Sometimes, it will be possible to find the particular solution without knowing this general solution. This method is known as the *method of undetermined coefficients*. It takes full advantage of the "forcing" function $f(t)$. The choice of $x_p(t)$ depends on $f(t)$. We use our experience from the method of undetermined coefficients in Chap. 3 and proceed as follows.

($\star$) If $f(t) = b$, where $b \in \mathbb{R}^n$ is a constant vector, we might reasonably expect a constant solution, say

$$x_p(t) = c.$$

Since $x_p'(t) = 0$ and $x_p(t)$ satisfies (5.14), we have to solve the linear system of algebraic equations

$$Ac = -b.$$

Example 5.18. Find the particular solution of

$$x'(t) = \begin{pmatrix} 1 & 0 \\ -1 & 3 \end{pmatrix} x(t) + \begin{pmatrix} 1 \\ 1 \end{pmatrix}.$$

Solution. Since $b = (1, 1)^T$ is a constant vector, we have to solve the system of linear algebraic equations

$$\begin{pmatrix} 1 & 0 \\ -1 & 3 \end{pmatrix} \begin{pmatrix} c_1 \\ c_2 \end{pmatrix} = - \begin{pmatrix} 1 \\ 1 \end{pmatrix}.$$

We obtain

$$c_1 = -1, \quad c_2 = -\frac{2}{3}, \quad \text{i.e.} \quad x_p(t) = - \begin{pmatrix} 1 \\ 2 \\ \frac{2}{3} \end{pmatrix}.$$

$\square$

(⋆) If $f(t) = e^{kt} b$, where k is not an eigenvalue of A and $b \in \mathbb{R}^n$, then

$$x_p(t) = e^{kt} c,$$

where $c \in \mathbb{R}^n$ is a constant vector.

Example 5.19. Find the particular solution of

$$x'(t) = \begin{pmatrix} 0 & 1 \\ 1 & 0 \end{pmatrix} x(t) + e^{2t} \begin{pmatrix} 0 \\ 1 \end{pmatrix}.$$

Solution. The eigenvalues ± 1 of A are different from $k = 2$. Thus, the particular solution has the form $x_p(t) = e^{2t} c$.
Substituting $x_p(t)$ and $x'_p(t) = 2e^{2t} c$ into the given system, we have

$$2e^{2t} c = e^{2t} \begin{pmatrix} 0 & 1 \\ 1 & 0 \end{pmatrix} c + e^{2t} \begin{pmatrix} 0 \\ 1 \end{pmatrix}, \quad \text{i.e.} \quad c = \frac{1}{3} \begin{pmatrix} 1 \\ 2 \end{pmatrix}.$$

Hence

$$x_p(t) = \frac{1}{3} e^{2t} \begin{pmatrix} 1 \\ 2 \end{pmatrix}.$$

$\square$

Remark 5.20. If k is a single eigenvalue of A, we cannot represent the solution in the form $x_p(t) = te^{kt}c$. Instead it is necessary to use the ansatz

$$x_p(t) = e^{kt}(a\,t + b),$$

where $a \in \mathbb{R}^n$ and $b \in \mathbb{R}^n$ are determined by substituting this ansatz into the given problem. Moreover, if k is a double eigenvalue of A, we choose

$$x_p(t) = e^{kt}(a\,t^2 + b\,t + c), \quad a, b, c \in \mathbb{R}^n.$$

$\square$

$(\star)$ If $f(t) = \cos(kt)b$ or $f(t) = \sin(kt)b$, in both cases, we solve the system for $f(t) = e^{ikt}b$ and compare real and imaginary parts with the given $f(t)$.

$(\star)$ If $f(t) = P_n(t)b$, where $P_n(t)$ is a polynomial of degree n, then

$$x_p(t) = Q_n(t)c, \quad c \in \mathbb{R}^n,$$

where $Q_n(t)$ is also a polynomial of degree n.

Example 5.21. Using the method of undetermined coefficients, find the particular solution of

$$x'(t) = \begin{pmatrix} 1 & 0 \\ 0 & -1 \end{pmatrix} x(t) + \cos(t) \begin{pmatrix} -1 \\ 1 \end{pmatrix}.$$

Solution. Instead of the given ODE, we consider the problem

$$y'(t) = \begin{pmatrix} 1 & 0 \\ 0 & -1 \end{pmatrix} y(t) + e^{it} \begin{pmatrix} -1 \\ 1 \end{pmatrix} \tag{5.24}$$

and assume that a particular solution of this modified problem has the form

$$y_p(t) = e^{it}c. \tag{5.25}$$

Substituting (5.25) into (5.24), we get

$$ie^{it}c = e^{it} \begin{pmatrix} 1 & 0 \\ 0 & -1 \end{pmatrix} c + e^{it} \begin{pmatrix} -1 \\ 1 \end{pmatrix}.$$

Hence,

$$i \begin{pmatrix} c_1 \\ c_2 \end{pmatrix} = \begin{pmatrix} c_1 \\ -c_2 \end{pmatrix} + \begin{pmatrix} -1 \\ 1 \end{pmatrix}.$$

It can be easily shown that the solution of this system is

$$c = \frac{1}{2} \begin{pmatrix} 1+i \\ 1-i \end{pmatrix}.$$

Substituting c into (5.25), we obtain

$$y_p(t) = \frac{1}{2} e^{it} \begin{pmatrix} 1+i \\ 1-i \end{pmatrix} = \frac{1}{2} (\cos(t) + i \sin(t)) \begin{pmatrix} 1+i \\ 1-i \end{pmatrix}$$

$$= \frac{1}{2} \begin{pmatrix} \cos(t) - \sin(t) + i(\cos(t) + \sin(t)) \\ \cos(t) + \sin(t) + i(\cos(t) - \sin(t)) \end{pmatrix}.$$

The desired particular solution $x_p(t)$ of the given problem is the real part of $y_p(t)$, that is,

$$x_p(t) = \frac{1}{2} \begin{pmatrix} \cos(t) - \sin(t) \\ \cos(t) + \sin(t) \end{pmatrix}.$$

□

The usefulness of the method of undetermined coefficients is further enhanced by the application of the *principle of superposition* which is presented in the following theorem.

Theorem 5.22. *If $x_1(t)$ and $x_2(t)$ are solutions of*

$$x_1'(t) = Ax_1(t) + f_1(t)$$

and

$$x_2'(t) = Ax_2(t) + f_2(t)$$

respectively, then

$$x_p(t) = c_1 x_1(t) + c_2 x_2(t)$$

is the solution of

$$x'(t) = Ax(t) + c_1 f_1(t) + c_2 f_2(t).$$

Example 5.23. Find a particular solution of

$$x'(t) = \begin{pmatrix} 1 & 0 \\ 0 & -1 \end{pmatrix} x(t) + (\cos(t) - 1) \begin{pmatrix} -1 \\ 1 \end{pmatrix}.$$

Solution. We set

$$f_1(t) \equiv \cos(t) \begin{pmatrix} -1 \\ 1 \end{pmatrix}, \quad f_2(t) \equiv \begin{pmatrix} 1 \\ -1 \end{pmatrix}.$$

For the system

$$x_1'(t) = \begin{pmatrix} 1 & 0 \\ 0 & -1 \end{pmatrix} x_1(t) + \cos(t) \begin{pmatrix} -1 \\ 1 \end{pmatrix}$$

we use the real part of $y_p(t)$ as computed in Example 5.21. So

$$x_{1,p}(t) = \frac{1}{2} \begin{pmatrix} \cos(t) - \sin(t) \\ \cos(t) + \sin(t) \end{pmatrix}.$$

For the system

$$x_2'(t) = \begin{pmatrix} 1 & 0 \\ 0 & -1 \end{pmatrix} x_2(t) + \begin{pmatrix} -1 \\ 1 \end{pmatrix}$$

we try $x_{2,p}(t) = c$, which leads to the particular solution

$$x_{2,p}(t) = \begin{pmatrix} -1 \\ -1 \end{pmatrix}.$$

Therefore, the particular solution of the given problem can be composed as the sum of $x_{1,p}(t)$ and $x_{p,2}(t)$, i.e.,

$$x_p(t) = \frac{1}{2} \begin{pmatrix} \cos(t) - \sin(t) \\ \cos(t) + \sin(t) \end{pmatrix} + \begin{pmatrix} -1 \\ -1 \end{pmatrix} = \frac{1}{2} \begin{pmatrix} \cos(t) - \sin(t) - 2 \\ \cos(t) + \sin(t) - 2 \end{pmatrix}.$$

$\square$

We observe that the procedure for choosing $x_p(t)$ is substantially the same as that given in Sect. 3.3. The method is applicable if the inhomogeneity $f(t)$ is an exponential, sinusoidal, or polynomial function or consists of sums or products of these.

At the end of this chapter, we present some theorems on the existence and uniqueness of solutions of initial value problems.

Theorem 5.24. *Let $A \in \mathbb{R}^{n \times n}$ be a constant matrix and $f(t) \in \mathbb{R}^n$ a continuous vector function on an interval I. Then the linear inhomogeneous initial value problem*

$$x'(t) = Ax(t) + f(t), \quad x(t_0) = c, \quad t_0 \in I, \quad c \in \mathbb{R}^n, \tag{5.26}$$

has a unique solution on I that is given by

$$x(t) = e^{(t-t_0)A} c + e^{tA} \int_{t_0}^{t} e^{-sA} f(s) ds. \tag{5.27}$$

Proof. See, e.g., [1].   ∎

In practice, the calculation of the exponential matrix in (5.27) is difficult even if the system of ODEs is a homogeneous one.

The following theorems are more general cases when A is not a constant matrix.

Theorem 5.25. *Let $A(t) \in \mathbb{R}^{n \times n}$ be a matrix function which is continuous on the interval I. Then the homogeneous initial value problem*

$$x'(t) = A(t)x(t), \quad x(t_0) = c, \quad t_0 \in I, \quad c \in \mathbb{R}^n,$$

has a unique solution on I.

Proof. See, e.g., [1].   ∎

Theorem 5.26. *Let $A(t) \in \mathbb{R}^{n \times n}$ and $f(t) \in \mathbb{R}^n$ be continuous functions on the interval I. Then the inhomogeneous initial value problem*

$$x'(t) = A(t)x(t) + f(t), \quad x(t_0) = c, \quad t_0 \in I, \quad c \in \mathbb{R}^n, \tag{5.28}$$

has a solution that is given by

$$x(t) = e^{A(t)-A(t_0)} c + e^{A(t)} \int_{t_0}^{t} e^{-A(s)} f(s) ds. \tag{5.29}$$

Proof. See, e.g., [1].   ∎

For the case that the system of ODEs contains nonlinear equations, we may use *Picard's iteration method*. That is, for a system of ODEs given by

$$x'(t) = F(t, x(t)), \quad x(t_0) = c,$$

the iterative relation

$$x_{k+1}(t) = c + \int_{t_0}^{t} F(s, x_k(s)) ds, \quad x_0(t) \equiv c.$$

can be used to approximate the solution $x(t)$.

5.6 Exercises

Exercise 5.1. Write the following scalar ODEs as a system of first-order ODEs:

(1) $y'' - 2y' - 3y = x^2$,   (2) $y'' + y = \cot(x)$,

(3) $y'' - 3y' + 2y = \cos(3x)$,   (4) $y'' - 4y = \sin(x)$,

(5) $y''' + p(t)y'' + q(x)y' + r(x)y = s(x)$,   (6) $y''' + 4y'' - 5y = 2x - 1$,

(7) $y^{(4)} + y = 2\cos(x) - \sin(2x)$,   (8) $(D^5 - 4D^2)y = 24$,

(9) $(D^3 - 1)y = 4 - 3x^2$,   (10) $D(D + 1)y = x$.

Exercise 5.2. Find the general solution of the following systems:

(1) $(D + 4)x + Dy = 1$, $(D - 2)x + y = t^2$,
(2) $(3D - 1)x + 4y = t$, $Dx - Dy = t - 1$,
(3) $(D^2 - 1)x + (D^2 - 2)y = -2\sin(t)$, $(D^2 + 1)x + D^2y = 0$,
(4) $D(x + y) + y = t$, $D^2x + Dy + x + y = t^2$,
(5) $(D + 1)(x + y) = 1$, $D^2x - Dy = 1/t$,
(6) $(2D + 1)x + (D^2 - 4)y = -7e^{-t}$, $Dx - (D + 2)y = -3e^{-t}$,
(7) $(D^3 + D^2 - 1)x + (D^3 + 2D^2 + 3D + 1)y = 3 - t$,
     $(D - 1)x + (D + 1)y = 3 - t$,
(8) $D^2x + (D - 1)y + (D + 2)z = 0$, $(D + 2)x + (2D - 3)y - (D - 6)z = 0$,
     $2Dx - (D + 3)y - Dz = 0$,
(9) $(D - 2)x - y - z = t$, $x + (2D - 1)y + 2z = 1$,
     $-3x + Dy + Dz = 0$,
(10) $(3D^2 - 1)x + Dy + D^2z = e^t$,
     $Dx - D^2y - (2D - 1)z = \cos(t) + t^2$,
     $(D - 1)x - D(D - 1)y - (D - 1)^2z = 3\sin(2t)$.

Exercise 5.3. Solve the following systems of equations by Laplace transforms:

(1) $x'' - 3x' - y' + 2y = 14t + 3$, $x' - 3x + y' = 1$,
     $x(0) = x'(0) = 0$, $y'(0) = 6$,
(2) $2(x' + x) + y' - y = 3t$, $x' + x + y' + y = 1$,
     $x(0) = 1$, $y(0) = 3$,
(3) $x' - 2x - y' - y = 6e^{3t}$, $2x' - 3x + y' - 3y = 6e^{3t}$,
     $x(0) = 1$, $y(0) = 0$,
(4) $x'' + 2x - y' = 2t + 5$, $x' - x + y' = -(2t + 1)$,
     $x(0) = 3$, $x'(0) = 0$, $y(0) = -3$,
(5) $x'' - x + 5y' = t$, $y'' - 4y - 2x' = -2$,
     $x(0) = x'(0) = 0$, $y(0) = 1$, $y'(0) = 0$,
(6) $x'' + x + y' = 2e^{-t}$, $y'' - 3y - x' = 0$,
     $x(0) = x'(0) = y(0) = y'(0) = 0$,
(7) $x'' + y' + 3x = 15e^{-t}$, $y'' - 4x' + 3y = 15\sin(2t)$,
     $x(0) = y'(0) = 0$, $x'(0) = y(0) = 1$,

(8) $x'' + y' - y = 0$, $\quad x' + 3x + y' - 4y + 3z = 0$, $\quad 2x' - x + z' - z = 0$,
$\quad x(0) = y(0) = z(0) = 0$, $\quad x'(0) = 1$,

(9) $x' + y' = y + z$, $\quad y' + z' = x + z$, $\quad x' + z' = x + y$,
$\quad x(0) = 2$, $\quad y(0) = -3$, $\quad z(0) = 1$,

(10) $x'' + y' + z = 0$, $\quad x' + x + y + z' = t$, $\quad y' + z' - x = 1$,
$\quad x(0) = y(0) = z(0) = 0$, $\quad x'(0) = 1$.

Exercise 5.4. Determine the general solution for each of the following systems of linear ODEs:

(1) $x'(t) = \begin{pmatrix} 4 & 1 \\ -4 & 8 \end{pmatrix} x(t) + e^{6t} \begin{pmatrix} 1 \\ 6t \end{pmatrix}$

(2) $x'(t) = \begin{pmatrix} 2 & -5 \\ 1 & -2 \end{pmatrix} x(t) + \begin{pmatrix} \csc(t) \\ \sec(t) \end{pmatrix}$,

(3) $x'(t) = \begin{pmatrix} 0 & 1 \\ -2 & 3 \end{pmatrix} x(t) + e^t \begin{pmatrix} 0 \\ 3 \end{pmatrix}$,

(4) $x'(t) = \begin{pmatrix} 0 & -2 \\ 4 & -1 \end{pmatrix} x(t) + \begin{pmatrix} 3 - e^t \sin(t) \\ 2e^{-t} \cos(t) \end{pmatrix}$,

(5) $x'(t) = \begin{pmatrix} 3 & 3 \\ -1 & -1 \end{pmatrix} x(t) + \begin{pmatrix} e^{-t} \\ e^{2t} \end{pmatrix}$,

(6) $x'(t) = \begin{pmatrix} 0 & 2 \\ -2/3 & 0 \end{pmatrix} x(t) + \begin{pmatrix} \cos(t) \\ -\sin(t)/3 \end{pmatrix}$,

(7) $x'(t) = \begin{pmatrix} 4 & -2 \\ 8 & -4 \end{pmatrix} x(t) + \begin{pmatrix} t^{-3} \\ -t^{-2} \end{pmatrix}$,

(8) $x'(t) = \begin{pmatrix} 2 & 1 \\ -4 & 2 \end{pmatrix} x(t) + e^{2t} \begin{pmatrix} t \\ -1 \end{pmatrix}$,

(9) $x'(t) = \begin{pmatrix} 2 & -5 \\ 1 & -2 \end{pmatrix} x(t) - \begin{pmatrix} \sin(t) \\ -t \end{pmatrix}$,

(10) $x'(t) = \begin{pmatrix} -4 & 2 \\ 2 & 1 \end{pmatrix} x(t) + e^{2t} \begin{pmatrix} t \\ 3 \end{pmatrix}$.

Exercise 5.5. Find a particular solution of the following systems of linear ODEs by the method of undetermined coefficients:

(1) $x'(t) = \begin{pmatrix} 2 & 1 \\ 1 & 2 \end{pmatrix} x(t) + t \begin{pmatrix} 2 \\ 1 \end{pmatrix}$,

(2) $x'(t) = \begin{pmatrix} 0 & 1 \\ 1 & 0 \end{pmatrix} x(t) + e^t \begin{pmatrix} 0 \\ 1 \end{pmatrix}$,

(3) $x'(t) = \begin{pmatrix} 1 & -2 \\ 0 & -1 \end{pmatrix} x(t) - e^{3t} \begin{pmatrix} 1 \\ 2 \end{pmatrix}$,

(4) $x'(t) = \begin{pmatrix} 0 & -1 \\ 1 & 0 \end{pmatrix} x(t) + \sin(t) \begin{pmatrix} 0 \\ 1 \end{pmatrix}$,

(5) $x'(t) = \begin{pmatrix} 1 & 2 \\ 1 & 0 \end{pmatrix} x(t) + (2e^t - 3\sin(t)) \begin{pmatrix} 0 \\ 1 \end{pmatrix}$.

Exercise 5.6. The following system of two ODEs governs the variation of a mechanical system consisting of two masses on two springs:

$$x'' = -x + 3(y - x),$$
$$3y'' = -3(y - x) - 3y,$$

where x and y are the displacements of the masses from their positions of static equilibrium. By choosing $x' = u$ and $y' = v$, construct a 4×4 coefficient matrix and then solve this new system.

Exercise 5.7. An electrical circuit is described by the following system of ODEs:

$$x'(t) = \begin{pmatrix} 0 & 2 \\ -1 & -3 \end{pmatrix} x(t) + V(t) \begin{pmatrix} 1 \\ 3 \end{pmatrix},$$

where x_1 is the current through the inductance, x_2 is the voltage drop across the capacitor, and $V(t)$ is the voltage supplied by the external source.

1. Determine the fundamental matrix for the homogeneous system corresponding to the given problem.
2. Let

$$V(t) = \begin{cases} 1, & 0 \le t \le 1 \\ 0, & t > 1 \end{cases}.$$

Determine the solution of the system which satisfies the initial condition $x(0) = \mathbf{0}$.

Reference

1. Hartman, P.: Ordinary Differential Equations. Birkhäuser Verlag, Boston/Basel/Stuttgart (1982)

Chapter 6
Power Series Solutions

6.1 Introduction

In Chap. 3, we have observed that it is not always possible to solve ODEs of the form

$$P(x)y'' + Q(x)y' + R(x)y = 0. \tag{6.1}$$

Although in some specific cases we could find a closed-form solution for (6.1), these cases are really the exception rather than the rule. There is not a general analytical method for finding the solution of (6.1). Thus, we must seek other means of expression for the solution of (6.1). One possibility is to use *approximate methods* which represent the solution of the ODE as an analytical series in the neighborhood of a given point. In this chapter we will study some of these analytical approximate methods, and in the last two chapters, we will discuss some numerical methods for initial and boundary value problems. For practical applications an implicit solution in terms of elementary functions is less useful than an approximation of the solution by an infinite series or a numerical method.

Thus, this chapter focuses on methods by which the solution of an ODE is represented as an infinite series, the so-called *power series solution*. These methods can also be used to solve linear higher-order ODEs. Therefore, in the next section, we will concentrate on finding a power series solution for Eq. (6.1) written in standard form and expanded about a point x_0, where x_0 has the property that $P(x)$, $Q(x)$, and $R(x)$ have convergent power series about x_0. In other words the method only works when the coefficients of the ODE are real analytical functions at $x = x_0$. In this circumstance we call x_0 an *ordinary point* of the ODE if $P(x_0) \neq 0$. Any point that is not an ordinary point of (6.1) is called a *singular point*. If $P(x)$ is not zero at $x = x_0$, then in an interval about this point, away from the nearest point that makes $P(x)$ zero, we can divide the relation (6.1) by $P(x)$ and write

$$y'' + p(x)y' + q(x)y = 0. \tag{6.2}$$

M. Hermann and M. Saravi, *A First Course in Ordinary Differential Equations: Analytical and Numerical Methods*, DOI 10.1007/978-81-322-1835-7_6, © Springer India 2014

We now show that we can expect a solution in form of a power series for (6.2) which contains two arbitrary constants c_1 and c_2.

Let us assume that $y = y(x)$ is a solution of (6.2) and $y(x_0) = a$, $y'(x_0) = b$. From (6.2), we get

$$y''(x) = -p(x)y'(x) - q(x)y(x).$$

Hence, $y''(x_0)$ may be found. We have

$$y'''(x) = -p(x)y''(x) - p'(x)y'(x) - q(x)y'(x) - q'(x)y(x).$$

Since $y(x_0)$, $y'(x_0)$, and $y''(x_0)$ are known, the value $y'''(x_0)$ can be obtained. Now, if we continue this process, then every $y^{(n)}(x_0)$ may be calculated and we get a series for the function $y(x)$. This argument for finding a solution of the ODE in form of a power series seems reasonable.

It is tempting to use the power series method to attack virtually any ODE that we encounter. But the method works best for linear equations.

We begin with a very brief review of the main facts about power series, which should be familiar from calculus course.

6.2   Review of Power Series

A power series is an infinite series of the form

$$\sum_{n=0}^{\infty} a_n (x - x_0)^n, \tag{6.3}$$

where a_n and x_0 are constants. This power series is *convergent* at x if

$$\lim_{N \to \infty} \sum_{n=0}^{N} a_n (x - x_0)^n$$

exists and is finite. Otherwise, it is *divergent*.

If for all x in the interval $|x - x_0| < R$ the power series (6.3) converges and is divergent whenever $|x - x_0| > R$, where $0 \le R \le \infty$, then R is called the *radius of convergence* of the power series. The radius of convergence R is given by

$$R = \lim_{n \to \infty} \left| \frac{a_n}{a_{n+1}} \right|.$$

Without loss of generality, we restrict our attention to power series of the form

$$\sum_{n=0}^{\infty} a_n x^n, \tag{6.4}$$

whose center is $x_0 = 0$. For the power series (6.4), precisely one of the following statements is true:

(1) $\sum_{n=0}^{\infty} a_n x^n$ converges only at $x = 0$.

(2) $\sum_{n=0}^{\infty} a_n x^n$ converges for all x.

(3) There exists a positive number R such that this series converges (absolutely) for $|x| < R$ and is divergent for $|x| > R$.

The cases $x = \pm R$ must be treated separately.

Suppose that $\sum_{n=0}^{\infty} a_n x^n$ has the radius of convergence R, and let

$$y(x) = \sum_{n=0}^{\infty} a_n x^n, \quad |x| < R.$$

Then the derivatives of the function $y(x)$ can be obtained term-wise on the interval $|x| < R$. Thus,

$$y'(x) = \sum_{n=0}^{\infty} n a_n x^{n-1} = \sum_{n=1}^{\infty} n a_n x^{n-1},$$

$$y''(x) = \sum_{n=0}^{\infty} n(n-1) a_n x^{n-2} = \sum_{n=2}^{\infty} n(n-1) a_n x^{n-2},$$

and so on.

Definition 6.1. A function defined on the interval I which contains the point $x = x_0$ is said to be *analytical* at x_0 if it can be represented by a convergent power series centered at $x = x_0$ with a positive radius of convergence. □

We can determine many other analytical functions using the following theorem.

Theorem 6.2. *If $f(x)$ and $g(x)$ are analytical at x_0, then this is also true for $f(x) \pm g(x)$, $f(x) \times g(x)$, and $f(x)/g(x)$ (provided $g(x_0) \neq 0$).*

In a calculus course, it is normally proved that if a function $f(x)$ is analytical at $x = x_0$, then the power series representation of that function will be unique and is given by

$$f(x) = \sum_{n=0}^{\infty} \frac{f^{(n)}(x_0)}{n!}(x - x_0)^n. \tag{6.5}$$

The representation (6.5) is called the *Taylor series* expansion of $f(x)$ about $x = x_0$. If $x_0 = 0$, then (6.5) is reduced to a series called the *Maclaurin series* expansion of $f(x)$, i.e.,

$$f(x) = \sum_{n=0}^{\infty} \frac{f^{(n)}(0)}{n!}x^n.$$

We give here the Maclaurin series expansion of some elementary functions that may be used later in this chapter or in the next chapter (see also Appendix A):

(1) $e^x = 1 + \dfrac{x}{1!} + \dfrac{x^2}{2!} + \dfrac{x^3}{3!} + \cdots, \quad -\infty < x < \infty.$

(2) $\cos(x) = 1 - \dfrac{x^2}{2!} + \dfrac{x^4}{4!} - \dfrac{x^6}{6!} + \cdots, \quad -\infty < x < \infty.$

(3) $\sin(x) = x - \dfrac{x^3}{3!} + \dfrac{x^5}{5!} - \dfrac{x^7}{7!} + \cdots, \quad -\infty < x < \infty.$

(4) $\tan^{-1}(x) = x - \dfrac{x^3}{3} + \dfrac{x^5}{5} - \dfrac{x^7}{7} + \cdots, \quad -1 \leq x \leq 1.$

(5) $(1 - x)^{-1} = 1 + x + x^2 + \cdots, \quad -1 < x < 1.$

(6) $\ln(1 + x) = x - \dfrac{x^2}{2} + \dfrac{x^3}{3} - \dfrac{x^4}{4} + \cdots, \quad -1 < x < 1.$

(7) $\ln(x) = (x - 1) - \dfrac{(x - 1)^2}{2} + \dfrac{(x - 1)^3}{3} - \dfrac{(x - 1)^4}{4} + \cdots, \quad 0 < x \leq 2.$

In the next two sections, we try to obtain power series solutions of those ODEs that have ordinary or singular points. To deal with the case that there exists an ordinary point is much easier than the one with a singular point. However, singular points often turn out to be the points of major interest in an applied problem. A number of ODEs such as Bessel's equation, Laguerre's equation, and many others contain singular points.

For the remainder of this chapter, we focus our attention on Eq. (6.1). Its singularities, if there are, can be subdivided into two classes, regular singularities and irregular singularities.

Definition 6.3. The point $x = x_0$ is called a *regular singular point* of Eq. (6.1) if it is a singular point and the following limits exist:

$$\lim_{x \to x_0} \frac{(x - x_0)Q(x)}{P(x)} \quad \text{and} \quad \lim_{x \to x_0} \frac{(x - x_0)^2 R(x)}{P(x)}.$$

Otherwise, it is called an *irregular singular point*.   □

Example 6.4. Determine whether $x_0 = 0$ and $x_0 = 1$ are ordinary points or regular singular points of Bessel's equation given by

$$x^2 y'' + xy' + (x^2 - \lambda^2)y = 0, \quad \lambda \geq 0.$$

Solution. Since $P(1) = 1 \neq 0$, the point $x_0 = 1$ is an ordinary point. For $x_0 = 0$, it holds $P(0) = 0$. Therefore, we have to study

$$\lim_{x \to x_0} \frac{(x - x_0)Q(x)}{P(x)} = \lim_{x \to x_0} \frac{(x - 0)x}{x^2} = 1,$$

$$\lim_{x \to x_0} \frac{(x - x_0)^2 R(x)}{P(x)} = \lim_{x \to x_0} \frac{(x - 0)^2(x^2 - \lambda^2)}{x^2} = -\lambda^2.$$

Looking at Definition 6.3, we see that $x_0 = 0$ is a regular singular point.

□

Example 6.5. Let $x_0 = \pm 1$. Determine the kind of singularity for the equation

$$(x^2 - 1)^2 y'' + 6(x - 1)y' + (x + 1)y = 0.$$

Solution. Obviously, it holds $P(\pm 1) = 0$. Thus, the points $x_0 = 1$ and $x_0 = -1$ are singular points. For $x = 1$, we have

$$\lim_{x \to x_0} \frac{(x - x_0)Q(x)}{P(x)} = \lim_{x \to 1} \frac{(x - 1)6(x - 1)}{(x^2 - 1)^2} = \frac{3}{2},$$

$$\lim_{x \to x_0} \frac{(x - x_0)^2 R(x)}{P(x)} = \lim_{x \to 1} \frac{(x - 1)^2(x + 1)}{(x^2 - 1)^2} = \frac{1}{2}.$$

Therefore, $x = 1$ is a regular singular point. But for $x = -1$, we obtain

$$\lim_{x \to x_0} \frac{(x - x_0)Q(x)}{P(x)} = \lim_{x \to -1} \frac{(x + 1)6(x - 1)}{(x^2 - 1)^2} = \infty.$$

Hence, this is an irregular singular point.

□

In the next sections, we will consider the solution of Eq. (6.1) with ordinary points and regular singular points. The answer to the basic question when the method of series substitution can be expected to work is given by the Fuchs's Theorem, which asserts that we can always obtain a power series solution, provided we expand the series about a point that is an ordinary point or at least a regular singular point. If we try to develop an expansion about an irregular or essential singularity, our method may fail. Fortunately, the more important equations of mathematical physics, such as Bessel's equation, Legendre's equation, Chebyshev's equation, and many others have no irregular singularities in the finite plane.

6.3 Series Solutions About an Ordinary Point

Let us assume that in (6.2) the coefficients $p(x)$ and $q(x)$ are analytical at $x = x_0$. That is, they can be expressed as power series about x_0 which are convergent for $|x - x_0| < r$. Hence, x_0 is an ordinary point. For $x_0 = 0$, we can formulate the following statement.

Theorem 6.6. *Assume that in (6.2) the coefficients $p(x)$ and $q(x)$ are analytical at $x = 0$. Then every solution of this ODE which satisfies the initial conditions $y(0) = a_0$ and $y'(0) = a_1$ can uniquely be expanded as a power series given by*

$$y(x) = \sum_{n=0}^{\infty} a_n x^n, \quad |x| < r, \tag{6.6}$$

where r is the corresponding radius of convergence.

Proof. Since $p(x)$ and $q(x)$ are analytical, we can write

$$p(x) = \sum_{n=0}^{\infty} A_n x^n = A_0 + A_1 x + A_2 x^2 + \cdots,$$

$$q(x) = \sum_{n=0}^{\infty} B_n x^n = B_0 + B_1 x + B_2 x^2 + \cdots,$$

where $|x| < r$. Now we try to find a solution of (6.2) in the form

$$y(x) = a_0 + a_1 x + a_2 x^2 + a_3 x^3 + a_4 x^4 + a_5 x^5 + \cdots. \tag{6.7}$$

We have

$$y'(x) = a_1 + 2a_2 x + 3a_3 x^2 + 4a_4 x^3 + 5a_5 x^4 + \cdots \tag{6.8}$$

and

$$y''(x) = 2a_2 + 6a_3 x + 12a_4 x^2 + 20a_5 x^3 + \cdots. \tag{6.9}$$

By substituting the series of $y(x)$, $y'(x)$, $y''(x)$, $p(x)$, and $q(x)$ into (6.2) and comparing both sides, we find $a_2, a_3, a_4, a_5, \cdots$ in terms of a_0 and a_1. Using the initial conditions, the series solution can uniquely be determined. ∎

We will illustrate the case where we have ordinary points by examples.

Example 6.7. Find the general solution of the linear ODE

$$y''(x) + \sinh(x)y'(x) + e^{-x}y(x) = 0.$$

Solution. The coefficients of this ODE are analytical at $x = 0$. Hence, there exists a series solution of the form (6.6). The first two derivatives of this series are given in (6.8) and (6.9).

It holds

$$\sinh(x) = x + \frac{x^3}{3!} + \frac{x^5}{5!} + \frac{x^7}{7!} + \cdots,$$

$$e^{-x} = 1 - \frac{x}{1!} + \frac{x^2}{2!} - \frac{x^3}{3!} + \cdots.$$

Substituting these relations and the series of y, y', and y'' into the given ODE, we obtain

$$(2a_2 + 6a_3x + 12a_4x^2 + 20a_5x^3 + \cdots)$$

$$+ (x + \frac{x^3}{3!} + \frac{x^5}{5!} + \frac{x^7}{7!} + \cdots)(a_1 + 2a_2x + 3a_3x^2 + 4a_4x^3 + 5a_5x^4 + \cdots)$$

$$+ (1 - \frac{x}{1!} + \frac{x^2}{2!} - \frac{x^3}{3!} + \cdots)(a_0 + a_1x + a_2x^2 + a_3x^3 + a_4x^4 + a_5x^5 + \cdots)$$

$$= 0.$$

Equating the coefficients of terms in the same power of x on both sides of this equation, we obtain

$$0 = 2a_2 + a_0, \text{ i.e. } a_2 = -\frac{a_0}{2},$$

$$0 = 6a_3 + 2a_2 - a_0, \text{ i.e. } a_3 = \frac{a_0}{6} - \frac{a_1}{3},$$

$$0 = 12a_4 + 3a_2 - a_1 + \frac{a_0}{2}, \text{ i.e. } a_4 = \frac{a_0}{12} + \frac{a_1}{12},$$

$$0 = 20a_5 + 4a_3 - a_2 + \frac{2a_1}{3} - \frac{a_0}{6}, \text{ i.e. } a_5 = -\frac{a_0}{15} - \frac{a_1}{15}.$$

Substituting these results into (6.7), we get

$$y(x) = a_0 \left(1 - \frac{x^2}{2} + \frac{x^3}{6} + \frac{x^4}{12} - \frac{x^5}{15} + \cdots \right)$$

$$+ a_1 \left(x - \frac{x^3}{3} + \frac{x^4}{12} - \frac{x^5}{15} + \cdots \right).$$

□

In finding a series solution of an ODE, it will usually not be easy to develop the general form of the series. However, if the equation has a closed-form solution, this solution can be found.

Upon collecting the coefficients of terms in the same power of x, a *recurrence relation* is obtained, and it can be shown that this relation determines all of the coefficients in terms of a_0 and a_1. To do this, after writing (6.4) in terms of all summations then by shifting their indices (if it is necessary), we obtain all summations over the same power of x, say x^k. Although we may have a tedious computation, this strategy is quite straightforward. For a clear understanding of this subject, let us consider some examples.

Example 6.8. In quantum mechanics the study of Schrödinger's equation for the case of a harmonic oscillator leads to the treatment of *Hermite's equation*

$$y'' - 2xy' + 2\lambda y = 0,$$

where λ is a parameter. Find by the power series method the general solution for $\lambda = 2$.

Solution. With $\lambda = 2$, we have $y'' - 2xy' + 4y = 0$. Let $y(x) = \sum_{n=0}^{\infty} a_n x^n$, then

$$y'(x) = \sum_{n=0}^{\infty} n a_n x^{n-1} \quad \text{and} \quad y''(x) = \sum_{n=0}^{\infty} n(n-1) a_n x^{n-2}.$$

Substituting these expressions into the given ODE, we obtain

$$\sum_{n=0}^{\infty} n(n-1) a_n x^{n-2} - 2x \sum_{n=0}^{\infty} n a_n x^{n-1} + 4 \sum_{n=0}^{\infty} a_n x^n = 0.$$

Simplifying this equation leads to

$$\sum_{n=0}^{\infty} n(n-1) a_n x^{n-2} - 2 \sum_{n=0}^{\infty} (n-2) a_n x^n = 0.$$

Now, we shift the index in the second series by lowering the index range by 2. We obtain

$$\sum_{n=0}^{\infty} n(n-1)a_n x^{n-2} - 2\sum_{n=2}^{\infty} (n-4)a_{n-2}x^{n-2} = 0.$$

For the first series, we can write

$$\sum_{n=0}^{\infty} n(n-1)a_n x^{n-2} = \sum_{n=2}^{\infty} n(n-1)a_n x^{n-2}.$$

Therefore, it follows

$$\sum_{n=2}^{\infty} n(n-1)a_n x^{n-2} - 2\sum_{n=2}^{\infty} (n-4)a_{n-2}x^{n-2} = 0.$$

We may write these two series under one symbol. We get

$$\sum_{n=2}^{\infty} \left(n(n-1)a_n - 2(n-4)a_{n-2} \right)x^{n-2} = 0.$$

Since $x^{n-2} \neq 0$, $n \geq 2$, each coefficient must vanish:

$$n(n-1)a_n - 2(n-4)a_{n-2} = 0, \quad n \geq 2.$$

This is a recurrence relation which can be useful to determine all $a_n \geq 2$. For this reason we write

$$a_n = \frac{2(n-4)a_{n-2}}{n(n-1)}, \quad n \geq 2.$$

For $n = 2$ we have $a_2 = -2a_0$ and for $n = 4$ we get $a_4 = 0$. Thus, for each even n, $n \geq 4$, it holds $a_n = 0$. On the other hand, when n is odd, i.e., $n = 2k + 1$, we obtain

$$a_{2k+1} = \frac{2(2k-3)a_{2k-1}}{(2k+1)(2k)}, \quad k \geq 1.$$

Substituting $k = 1, 2, \ldots$, respectively, it can be shown that

$$a_{2k+1} = \frac{2^k(-3)(-1)(1)(3)\cdots(2k-3)a_1}{(2k+1)!}, \quad k \geq 1.$$

Now, we can write the general solution in terms of a_0 and a_1 as follows:

$$y(x) = \sum_{n=0}^{\infty} a_n x^n = a_0 + a_1 x + \sum_{k=1}^{\infty} a_{2k} x^{2k} + \sum_{k=1}^{\infty} a_{2k+1} x^{2k+1}$$

$$= a_0 + a_1 x + a_2 x^2 + \sum_{k=1}^{\infty} a_{2k+1} x^{2k+1}$$

$$= a_0 + a_1 x - 2a_0 x^2 + \sum_{k=1}^{\infty} \frac{2^k (-3)(-1)(1)(3) \cdots (2k-3) a_1}{(2k+1)!} x^{2k+1}$$

$$= a_0 (1 - 2x^2) + a_1 \left(x + \sum_{k=1}^{\infty} \frac{2^k (-3)(-1)(1)(3) \cdots (2k-3)}{(2k+1)!} x^{2k+1} \right).$$

For the special choice $a_0 = 1$ and $a_1 = 0$, a closed-form solution results, namely, $y(x) = 1 - 2x^2$.

It should be pointed out that one may write

$$a_{2k+1} = \frac{2(2k-3)a_{2k-1}}{(2k+1)(2k)} = \frac{(2k-3)a_{2k-1}}{(2k+1)(k)}$$

$$= \frac{(-3)(-1)(1)(3) \cdots (2k-3)a_1}{\big((3)(5) \cdots (2k+1)\big)\big((1)(2) \cdots (k)\big)}$$

$$= \frac{3a_1}{(2k-3)(4k^2-1)(k!)}.$$

Therefore, the general solution can be written in the simpler form

$$y(x) = a_0 \left(1 - 2x^2 \right) + a_1 \left(x + 3 \sum_{k=1}^{\infty} \frac{x^{2k+1}}{(2k-3)(4k^2-1)(k!)} \right).$$

□

Example 6.9. Find a power series solution of the initial value problem

$$(1 + x^2)y'' - 4xy' + 6y = 0, \quad y(0) = 1, \quad y'(0) = 2.$$

Solution. We observe that all points except $x = \pm i$ are ordinary points for this equation. Thus, we can assume that the solution has the form (6.6) for any $x_0 \neq \pm i$. As in Example 6.8, we have

$$y(x) = \sum_{n=0}^{\infty} a_n x^n, \quad y'(x) = \sum_{n=0}^{\infty} n a_n x^{n-1} \quad \text{and} \quad y''(x) = \sum_{n=0}^{\infty} n(n-1) a_n x^{n-2}.$$

Substituting these expressions into the given ODE leads to

$$(1 + x^2) \sum_{n=0}^{\infty} n(n-1) a_n x^{n-2} - 4x \sum_{n=0}^{\infty} n a_n x^{n-1} + 6 \sum_{n=0}^{\infty} a_n x^n = 0.$$

Simplifying this equation gives

$$\sum_{n=0}^{\infty} n(n-1) a_n x^{n-2} + \sum_{n=0}^{\infty} n(n-1) a_n x^n - 4 \sum_{n=0}^{\infty} n a_n x^n + 6 \sum_{n=0}^{\infty} a_n x^n = 0.$$

Writing the second, third, and fourth series under one symbol, we obtain

$$\sum_{n=0}^{\infty} n(n-1) a_n x^{n-2} + \sum_{n=0}^{\infty} (n-2)(n-3) a_n x^n = 0.$$

Re-indexing the second series by 2 and considering that the first two terms of the first series vanish, we get

$$\sum_{n=2}^{\infty} n(n-1) a_n x^{n-2} + \sum_{n=2}^{\infty} (n-4)(n-5) a_{n-2} x^{n-2} = 0.$$

Now, these two series can be written under one symbol and we obtain

$$\sum_{n=2}^{\infty} \left(n(n-1) a_n + (n-4)(n-5) a_{n-2} \right) x^{n-2} = 0.$$

This relation leads to

$$n(n-1) a_n + (n-4)(n-5) a_{n-2} = 0,$$

i.e.,

$$a_n = -\frac{(n-4)(n-5) a_{n-2}}{n(n-1)}, \quad n \geq 2.$$

For $n = 2$ we have $a_2 = -3a_0$ and for $n = 3$ we get $a_3 = -a_1/3$. If we put $n = 4$, we obtain $a_4 = 0$, hence $a_4 = a_6 = a_8 = \cdots = 0$. It is clear that for $n = 5$, we have $a_5 = 0$. Thus, $a_5 = a_7 = a_9 = \cdots = 0$. This yields

$$y(x) = a_0 + a_1 x + a_2 x^2 + a_3 x^3 = a_0 + a_1 x - 3a_0 x^2 - \frac{1}{3} a_1 x^3$$

$$= a_0(1 - 3x^2) + a_1 \left(x - \frac{1}{3} x^3 \right).$$

Applying the given initial conditions to this solution, we obtain $a_0 = 1$ and $a_1 = 2$. Hence, the solution of the initial value problem is

$$y(x) = 1 - 3x^2 + 2\left(x - \frac{1}{3}x^3\right).$$

$\square$

6.4   Series Solutions About a Regular Singular Point

Let $x = 0$ be a regular singular point of (6.2), where $p(x)$ and $q(x)$ are rational functions. Then by definition, $p(x)$ cannot contain a factor of x with exponent more than one in its denominator. That is,

$$p(x) = \frac{A_0}{x} + A_1 + A_2x + A_3x^2 + \cdots,$$

and similarly, $q(x)$ cannot have a factor of x with exponent more than two in its denominator. That is,

$$q(x) = \frac{B_0}{x^2} + \frac{B_1}{x} + B_2 + B_3x + \cdots.$$

Using these two results, we can expect that the solution of the ODE has the form

$$y(x) = \sum_{n=0}^{\infty} a_n x^{n+r} = a_0 x^r + a_1 x^{1+r} + a_2 x^{2+r} + \cdots. \tag{6.10}$$

If we substitute these series of $y(x)$, $y'(x)$, $y''(x)$, $p(x)$, and $q(x)$ into (6.10) and simplify, we obtain

$$0 = \left(r(r-1) + A_0r + B_0\right) a_0 x^{r-2}$$

$$+ \left((A_1r + B_1) a_0 + ((1+r)(r + A_0) + B_0)\right) a_1 x^{r-1} + \dots$$

The coefficient of x^{r-2} must vanish, i.e., we have

$$\left(r(r-1) + A_0r + B_0\right) a_0 = 0.$$

We assume that $a_0 \neq 0$; otherwise, we obtain the trivial solution $y(x) \equiv 0$. Thus,

$$r(r-1) + A_0r + B_0 = 0, \text{ i.e. } r^2 + (A_0 - 1)r + B_0 = 0.$$

This is a second-degree equation in terms of r and is called the *indicial equation* of (6.2). It is clear that this equation has two roots $r = r_1$ and $r = r_2$ which are called the *indicial roots* or *exponents* of the singularity. These roots may be real or complex, but in both cases, the real parts are of our interest.

We usually distinguish three cases corresponding to the nature of these roots. For the sake of simplicity, let us suppose that r_1 and r_2 are real. Then the following statement can be proved.

Theorem 6.10. *Suppose $x = 0$ is a singular ordinary point of (6.2), where $xp(x)$ and $x^2q(x)$ have Taylor expansions for all x in the interval $|x| < R$. Then*

1. *If the roots r_1 and r_2 of the indicial equation are distinct and do not differ by an integer, then there exist two linearly independent solutions of (6.2) of the form*

$$y_1(x) = \sum_{n=0}^{\infty} a_n x^{n+r_1}, \quad a_0 \neq 0,$$

$$y_2(x) = \sum_{n=0}^{\infty} b_n x^{n+r_2}, \quad b_0 \neq 0.$$

2. *If $r_1 = r_2$, then there always exist two linearly independent solutions of (6.2) of the form*

$$y_1(x) = \sum_{n=0}^{\infty} a_n x^{n+r_1}, \quad a_0 \neq 0,$$

$$y_2(x) = y_1 \ln(x) + \sum_{n=1}^{\infty} b_n x^{n+r_2}.$$

3. *If $r_1 - r_2 = N$, where N is a positive integer, then there always exist two linearly independent solutions of (6.2) of the form*

$$y_1(x) = \sum_{n=0}^{\infty} a_n x^{n+r_1}, \quad a_0 \neq 0,$$

$$y_2(x) = c\, y_1 \ln(x) + \sum_{n=0}^{\infty} b_n x^{n+r_2}, \quad b_0 \neq 0,$$

where c is a constant that could be zero.

Proof. See, e.g., [1].   ∎

Now we present for each of the three cases which have distinguished in Theorem 6.10 some examples.

Case 1. Suppose that the roots of the indicial equation are distinct and their difference is not an integer. We substitute r_1 into the recurrence relation. This enables us to find a_n in terms of a_0. Following the same procedure, using the second root r_2, we obtain the value of b_n in terms of b_0. We note that if r_1 and r_2 are non-integers but their difference is an integer, we may have a logarithm term. The reason of this situation will be clear when we discuss Cases 2 and 3.

Example 6.11. Find a series solution of

$$3xy'' - 2y' - y = 0.$$

Solution. The point $x = 0$ is a regular singular point for this equation. Hence, we have a solution of the form (6.6), and we obtain

$$y'(x) = \sum_{n=0}^{\infty}(n+r)a_n x^{n+r-1} \quad \text{and} \quad y''(x) = \sum_{n=0}^{\infty}(n+r)(n+r-1)a_n x^{n+r-2}.$$

Substituting these expressions into the given ODE leads to

$$0 = \sum_{n=0}^{\infty} 3(n+r)(n+r-1)a_n x^{n+r-1}$$

$$+ \sum_{n=0}^{\infty} 2(n+r)a_n x^{n+r-1} - \sum_{n=0}^{\infty} a_n x^{n+r}.$$

We write the first and the second series under one symbol and obtain

$$0 = \sum_{n=0}^{\infty}(n+r)(3n+3r-1)a_n x^{n+r-1} - \sum_{n=0}^{\infty} a_n x^{n+r}.$$

Re-indexing the second series by 1 and writing separately the term for $n = 0$ of the first series, we get

$$0 = r(3r-1)a_0 x^{r-1} + \sum_{n=1}^{\infty}(n+r)(3n+3r-1)a_n x^{n+r-1}$$

$$- \sum_{n=1}^{\infty} a_{n-1} x^{n+r-1}.$$

Writing these two series under one symbol, we obtain

$$0 = r(3r-1)a_0 x^{r-1} + \sum_{n=1}^{\infty} \left((n+r)(3n+3r-1)a_n - a_{n-1}\right)x^{n+r-1}.$$

The assumption $a_0 \neq 0$ implies $r(3r - 1) = 0$. That is, $r_1 = 1/3$ and $r_2 = 0$.
We also have

$$(n + r)(3n + 3r - 1)a_n - a_{n-1} = 0.$$

Thus,

$$a_n = \frac{a_{n-1}}{(n + r)(3n + 3r - 1)}, \quad n \geq 1. \tag{6.11}$$

Following the method outlined above, we substitute the root $r_1 = 1/3$ into the
right-hand side of (6.11) and determine successively $a_1, a_2, a_3, \ldots$ in terms of a_0.
For $r_1 = 1/3$, formula (6.11) reads

$$a_n = \frac{a_{n-1}}{n(3n + 1)}, \quad n \geq 1.$$

Starting with an arbitrary a_0, this recursion leads to

$$a_1 = \frac{a_0}{1(4)}, \quad a_2 = \frac{a_1}{2(7)} = \frac{a_0}{1 \times 2 \times (4 \times 7)},$$

$$a_3 = \frac{a_2}{3(10)} = \frac{a_0}{1 \times 2 \times 3 \times (4 \times 7 \times 10)}, \ldots$$

Thus,

$$a_n = \frac{a_0}{1 \times 2 \times 3 \times \cdots \times n \times (4 \times 7 \times 10 \times \cdots \times (3n + 1))}$$

$$= \frac{a_0}{n!(4 \times 7 \times 10 \times \cdots \times (3n + 1))}.$$

Therefore, for $r_1 = 1/3$ and the above determined values $a_1, a_2, \ldots$, we obtain

$$y_1(x) = \sum_{n=0}^{\infty} a_n x^{n + \frac{1}{3}} = a_0 x^{\frac{1}{3}} + \sum_{n=1}^{\infty} a_n x^{n + \frac{1}{3}}$$

$$= a_0 x^{\frac{1}{3}} + \sum_{n=1}^{\infty} \frac{a_0 x^{n + \frac{1}{3}}}{n!(4 \times 7 \times 10 \times \cdots \times (3n + 1))}$$

$$= a_0 x^{\frac{1}{3}} \left(1 + \sum_{n=1}^{\infty} \frac{x^n}{n!(4 \times 7 \times 10 \times \cdots \times (3n + 1))} \right).$$

To obtain a second solution, we use the root $r_2 = 0$ and proceed as above. We have

$$b_n = \frac{b_{n-1}}{n(3n-1)}, \quad n \geq 1.$$

That is,

$$b_1 = \frac{b_0}{1(2)}, \quad b_2 = \frac{b_1}{2(5)} = \frac{b_0}{1 \times 2 \times (2 \times 5)},$$

$$b_3 = \frac{b_2}{3(8)} = \frac{b_0}{1 \times 2 \times 3 \times (2 \times 5 \times 8)}, \cdots.$$

The general form of the b_n, $n \geq 1$, is

$$b_n = \frac{b_0}{1 \times 2 \times 3 \times \cdots \times n \times (2 \times 5 \times 8 \times \cdots \times (3n-1))}$$

$$= \frac{b_0}{n!(2 \times 5 \times 8 \times \cdots \times (3n-1))}.$$

Thus,

$$y_2(x) = \sum_{n=0}^{\infty} b_n x^n = b_0 + \sum_{n=1}^{\infty} b_n x^n$$

$$= b_0 + \sum_{n=1}^{\infty} \frac{b_0 x^n}{n!(2 \times 5 \times 8 \times \cdots \times (3n-1))}$$

$$= b_0 \left(1 + \sum_{n=1}^{\infty} \frac{x^n}{n!(2 \times 5 \times 8 \times \cdots \times (3n-1))} \right).$$

Note that the constants a_0 and b_0 play the same role as c_1 and c_2 in the general solution of a linear ODE.

□

Example 6.12. The ODE

$$4x^2 y'' + 4xy' + (4x^2 - 1)y = 0$$

has a regular singular point at $x = 0$. Find the general solution of the equation. Note, this ODE is a Bessel's equation with $\lambda = 1/2$.

Solution. For the solution we make the following ansatz

$$y(x) = \sum_{n=0}^{\infty} a_n x^{n+r}.$$

Then $y'(x)$ and $y''(x)$ are the same as given in the previous example. Substituting these expressions into the given ODE yields

$$0 = \sum_{n=0}^{\infty} 4(n+r)(n+r-1)a_n x^{n+r} + \sum_{n=0}^{\infty} 4(n+r)a_n x^{n+r}$$

$$+ \sum_{n=0}^{\infty} 4a_n x^{n+r+2} - \sum_{n=0}^{\infty} a_n x^{n+r}.$$

We can write the first, the second, and the fourth series under one symbol. We get

$$0 = \sum_{n=0}^{\infty} \left(4(n+r)^2 - 1\right)a_n x^{n+r} + \sum_{n=0}^{\infty} 4a_n x^{n+r+2}.$$

Shifting the index in the second series by 2 and writing the first two terms of the first series separately, we obtain

$$0 = (4r^2 - 1)a_0 x^r + \left(4(1+r)^2 - 1\right)a_1 x^{1+r}$$

$$+ \sum_{n=0}^{\infty} \left((4(n+r)^2 - 1)a_n + 4a_{n-2}\right) x^{n+r}. \tag{6.12}$$

The assumption $a_0 \neq 0$ implies $4r^2 - 1 = 0$. The corresponding roots are $r_1 = -1/2$ and $r_2 = 1/2$. We also have

$$\left(4(n+r)^2 - 1\right)a_n + 4a_{n-2} = 0.$$

Hence,

$$a_n = -\frac{4a_{n-2}}{4(n+r)^2 - 1}, \quad n \geq 2. \tag{6.13}$$

At the moment we do not distinguish whether $a_1 = 0$ or $a_1 \neq 0$.

Let us choose $r = 1/2$. Then the factor $\left(4(1+r)^2 - 1\right)$ in the second term of (6.12) does not vanish. Therefore, a_1 must be zero. Now, the recursion (6.13) implies $a_1 = a_3 = a_5 = \cdots = 0$.

But if n is even, i.e., $n = 2k$, formula (6.13) reads

$$a_{2k} = -\frac{4a_{2k-2}}{4\left(2k + \frac{1}{2}\right)^2 - 1} = -\frac{a_{2k-2}}{2k(2k + 1)}, \quad k \geq 1.$$

It is not difficult to show that

$$a_{2k} = \frac{(-1)^k a_0}{(2k + 1)!}, \quad k \geq 1.$$

Thus,

$$y_1 = \sum_{n=0}^{\infty} a_n x^{n+\frac{1}{2}} = a_0 x^{\frac{1}{2}} + \sum_{k=1}^{\infty} a_{2k} x^{2k+\frac{1}{2}}$$

$$= a_0 x^{\frac{1}{2}} + \sum_{k=1}^{\infty} \frac{(-1)^k a_0 x^{2k+\frac{1}{2}}}{(2k + 1)!}.$$

One may write

$$y_1(x) = a_0 x^{-\frac{1}{2}} \left(x + \sum_{k=1}^{\infty} \frac{(-1)^k x^{2k+1}}{(2k + 1)!}\right)$$

$$= a_0 x^{-\frac{1}{2}} \sin(x)$$

$$= a_0 \frac{\sin(x)}{\sqrt{x}}.$$

Now, if we choose $r = -1/2$, then it is not necessary that $a_1 = 0$. We have

$$a_n = -\frac{a_{n-2}}{(n - 1)n}, \quad n \geq 2.$$

If n is even, i.e., $n = 2k$, then it is not difficult to show that

$$a_{2k} = -\frac{a_{2k-2}}{(2k - 1)2k} = \frac{(-1)^k a_0}{(2k)!}, \quad k \geq 1.$$

If n is odd, i.e., $n = 2k + 1$, it can be shown

$$a_{2k+1} = -\frac{a_{2k-1}}{2k(2k + 1)} = \frac{(-1)^k a_1}{(2k + 1)!}, \quad k \geq 1.$$

Thus,

$$y_2(x) = \sum_{n=0}^{\infty} a_n x^{n-\frac{1}{2}}$$

$$= a_0 x^{-\frac{1}{2}} + a_1 x^{r-\frac{1}{2}} + \sum_{k=1}^{\infty} a_{2k} x^{2k-\frac{1}{2}} + \sum_{k=1}^{\infty} a_{2k+1} x^{2k+\frac{1}{2}}$$

$$= a_0 x^{-\frac{1}{2}} + a_1 x^{r-\frac{1}{2}} + \sum_{k=1}^{\infty} \frac{(-1)^k a_0}{(2k)!} x^{2k-\frac{1}{2}} + \sum_{k=1}^{\infty} \frac{(-1)^k a_1}{(2k+1)!} x^{2k-\frac{1}{2}}$$

$$= a_0 x^{-\frac{1}{2}} \left(1 + \sum_{k=1}^{\infty} \frac{(-1)^k x^{2k}}{(2k)!} \right) + a_1 x^{-\frac{1}{2}} \left(x + \sum_{k=1}^{\infty} \frac{(-1)^k x^{2k+1}}{(2k+1)!} \right)$$

$$= a_0 x^{-\frac{1}{2}} \cos(x) + a_1 x^{-\frac{1}{2}} \sin(x)$$

$$= a_0 \frac{\cos(x)}{\sqrt{x}} + a_1 \frac{\sin(x)}{\sqrt{x}}.$$

Since the second term in y_2 contains y_1, we can accept y_2 as the general solution. Note, the roots differ by the integer 1, but the general solution does not contain a logarithm term. The smaller root alone will give, in this case, two independent solutions.

□

Case 2. If the indicial equation has a double root, it is evident that only one series solution can be obtained. For the second linearly independent solution, we can use the order reduction formula if the first solution has a closed-form representation. Otherwise, we use the formula given in Theorem 6.10. In the following example, we try to find a second series solution.

Example 6.13. Find the general solution of the ODE

$$x^2 y'' + x(x - 1)y' + y = 0.$$

Solution. This equation has a regular singular point at $x = 0$. Hence, we choose the representation

$$y(x) = \sum_{n=0}^{\infty} a_n x^{n+r}.$$

We have

$$y'(x) = \sum_{n=0}^{\infty} (n+r)a_n x^{n+r-1}$$

$$y''(x) = \sum_{n=0}^{\infty} (n+r)(n+r-1)a_n x^{n+r-2}.$$

Substituting the series for y, y', and y'' into the given equation, we obtain

$$0 = \sum_{n=0}^{\infty} (n+r)(n+r-1)a_n x^{n+r} + \sum_{n=0}^{\infty} (n+r)a_n x^{n+r+1}$$

$$- \sum_{n=0}^{\infty} (n+r)a_n x^{n+r} + \sum_{n=0}^{\infty} a_n x^{n+r}.$$

We write the first, third, and fourth series with one symbol and get

$$0 = \sum_{n=0}^{\infty} (n+r-1)^2 a_n x^{n+r} + \sum_{n=0}^{\infty} (n+r)a_n x^{n+r+1}.$$

Changing the summation index of the second series and writing separately the first term of the first series yield

$$0 = (r-1)^2 a_0 x^r + \sum_{n=1}^{\infty} \left((n+r-1)^2 a_n + (n+r-1)a_{n-1} \right) x^{n+r}.$$

Since $a_0 \neq 0$, we have $(r-1)^2 = 0$. That is, $r_1 = r_2 = 1$, and we also have

$$(n+r-1)^2 a_n + (n+r-1)a_{n-1} = 0, \quad n \geq 1.$$

For $n \geq 1$, it holds $(n+r-1) \neq 0$; hence,

$$a_n = -\frac{a_{n-1}}{n+r-1}, \quad n \geq 1.$$

If we set $r = 1$, we obtain

$$a_n = -\frac{a_{n-1}}{n}, \quad n \geq 1.$$

It is clear that from this recurrence relation we will have

$$a_n = \frac{(-1)^n a_0}{n!}, \quad n \geq 1.$$

Substituting the value $r = 1$ and the above values of a_n, $n \geq 1$, into the series representation of $y(x)$, we come to

$$y_1(x) = \sum_{n=0}^{\infty} a_n x^{n+1} = a_0 x + \sum_{n=1}^{\infty} a_n x^{n+1} = a_0 x + \sum_{n=1}^{\infty} \frac{(-1)^n a_0 x^{n+1}}{n!}.$$

Obviously,

$$y_1(x) = a_0 x \left(1 + \sum_{n=1}^{\infty} \frac{(-1)^n x^n}{n!} \right) = a_0 x \, e^{-x}.$$

Thus, for the given ODE, the solution $y_1(x)$ can be written in a closed form, and we can use the method of order reduction to find the second linearly independent solution $y_2(x)$. We have $y_2(x) = u(x) y_1(x)$, where

$$u(x) = \int \frac{e^{-\int p dx}}{y_1^2} dx = \int \frac{e^x}{x} dx = \int \left(\frac{1}{x} + \frac{1}{1!} + \frac{x}{2!} + \frac{x^2}{3!} + \cdots \right) dx$$

$$= \ln(x) + \frac{x}{1! \, 1} + \frac{x^2}{2! \, 2} + \frac{x^3}{3! \, 3} + \cdots .$$

Then

$$y_2(x) = u(x) y_1(x) = x e^{-x} \left(\ln(x) + \frac{x}{1! \, 1} + \frac{x^2}{2! \, 2} + \frac{x^3}{3! \, 3} + \cdots \right).$$

$\square$

Remark 6.14. If $y_1(x)$ cannot be written as a closed-form solution, then by the indicial equation, we have

$$\mathcal{L}[y(x,r)] = (r - c)^2 x^r, \tag{6.14}$$

where $\mathcal{L}[y_1(x)] = 0$. Since in the right-hand side of (6.14) we have a multiple factor, its derivative also contains this factor with one lower power. Differentiating (6.14) gives

$$\frac{\partial}{\partial r} \mathcal{L}[y(x,r)] = \mathcal{L}\left[\frac{\partial y(x,r)}{\partial r} \right] = 2(r - c) x^r + (r - c)^2 x^r \ln(x). \tag{6.15}$$

Since for $r = c$ the right-hand sides of Eqs. (6.14) and (6.15) vanish, it holds

$$y_1(x) = y(x,r) = y(x,c), \quad y_2(x) = \left[\frac{\partial y(x,r)}{\partial r} \right]_{r=c}.$$

Therefore, we have only to determine $\dfrac{\partial y(x,r)}{\partial r}$. We have

$$y(x,r) = x^r + \sum_{n=1}^{\infty} a_n(r)x^{n+r}.$$

Thus,

$$\frac{\partial y(x,r)}{\partial r} = x^r \ln(x) + \sum_{n=1}^{\infty} a_n(r)x^{n+r} \ln(x) + \sum_{n=1}^{\infty} \frac{da_n}{dr} x^{n+r}$$

$$= y(x,r) \ln(x) + \sum_{n=1}^{\infty} \frac{da_n}{dr} x^{n+r}.$$

$\square$

Example 6.15. Find the solutions of the ODE

$$xy'' + (1-x)y' - y = 0.$$

Solution. The point $x = 0$ is a regular singular point for this equation. Hence, we determine the solution in the form

$$y(x) = \sum_{n=0}^{\infty} a_n x^{n+r}.$$

We have

$$y'(x) = \sum_{n=0}^{\infty} (n+r)a_n x^{n+r-1}$$

$$y''(x) = \sum_{n=0}^{\infty} (n+r)(n+r-1)a_n x^{n+r-2}.$$

Substituting these series into the given ODE leads to

$$0 = \sum_{n=0}^{\infty} (n+r)(n+r-1)a_n x^{n+r-1} + \sum_{n=0}^{\infty} (n+r)a_n x^{n+r-1}$$

$$- \sum_{n=0}^{\infty} (n+r)a_n x^{n+r} - \sum_{n=0}^{\infty} a_n x^{n+r}.$$

We write the first and the second series with one symbol and the same for the third and the fourth series, and we obtain

$$0 = \sum_{n=0}^{\infty} (n + r)^2 a_n x^{n+r-1} - \sum_{n=0}^{\infty} (n + r + 1) a_n x^{n+r}.$$

Re-indexing the second series by 1 and choosing separately $n = 0$ for the first series, we get

$$0 = r^2 a_0 x^{r-1} + \sum_{n=1}^{\infty} (n + r)^2 a_n x^{n+r-1} - \sum_{n=1}^{\infty} (n + r) a_{n-1} x^{n+r-1}.$$

Writing these two series as one series, we obtain

$$0 = r^2 a_0 x^{r-1} + \sum_{n=1}^{\infty} \left((n + r) a_n - a_{n-1} \right) (n + r) x^{n+r-1}.$$

Suppose that $a_0 \neq 0$, then $r^2 = 0$. That is, $r_1 = r_2 = 0$.
    We also have

$$\left((n + r) a_n - a_{n-1} \right)(n + r) = 0.$$

Since $n + r \neq 0$, it holds

$$a_n = \frac{a_{n-1}}{n + r}, \quad n \geq 1.$$

Substitute the root $r = 0$ and solve each for $a_1, \ a_2, \ a_3, \ldots$ in terms of a_0. We have

$$a_n = \frac{a_{n-1}}{n}, \quad n \geq 1.$$

That is,

$$a_n = \frac{a_0}{n!}, \quad n \geq 1.$$

Therefore, for $r = 0$ and the above determined values a_n, we obtain

$$y_1(x) = \sum_{n=0}^{\infty} a_n x^n = a_0 + \sum_{n=1}^{\infty} a_n x^n = a_0 + \sum_{n=1}^{\infty} \frac{a_0}{n!} x^n = a_0 e^x.$$

For finding the second solution $y_2(x)$, first, we use the procedure given in Remark 6.14. If we put $n = 1, 2, \ldots$, respectively, in the recurrence relation, we come to

$$a_n = \frac{a_0}{(1+r)(2+r)\cdots(n+r)}, \quad n \geq 1.$$

To find the derivative da_n/dr, we can use the logarithm and write

$$\ln(a_n) = \ln \left(\frac{a_0}{(1+r)(2+r)\cdots(n+r)} \right)$$

$$= \ln(a_0) - \ln \left((1+r)(2+r)\cdots(n+r) \right)$$

$$= \ln(a_0) - \left(\ln(1+r) + \ln(2+r) + \cdots + \ln(n+r) \right).$$

Differentiating with respect to r leads to

$$\frac{a_n'}{a_n} = -\left(\frac{1}{1+r} + \frac{1}{2+r} + \cdots + \frac{1}{n+r} \right).$$

That is,

$$a_n' = -a_n \left(\frac{1}{1+r} + \frac{1}{2+r} + \cdots + \frac{1}{n+r} \right).$$

Now, we set $r = 0$, and we obtain

$$a_n' = -\frac{a_0}{n!} \left(\frac{1}{1} + \frac{1}{2} + \cdots + \frac{1}{n} \right) = -\frac{a_0}{n!} H_n,$$

where

$$H_n \equiv \frac{1}{1} + \frac{1}{2} + \cdots + \frac{1}{n}.$$

Therefore,

$$y_2(x) = y_1(x) \ln(x) - a_0 \sum_{n=0}^{\infty} \frac{H_n}{n!} x^n.$$

$\square$

Case 3. If the two roots of the indicial equation differ by a nonzero integer, then we are encountered with a solution that may have a logarithmic term or may not. We demonstrate these two cases by two examples.

Example 6.16. Find two series solutions for the ODE

$$xy'' - (x + 2)y' + y = 0.$$

Solution. We start with the ansatz $y(x) = \sum_{n=0}^{\infty} a_n x^{n+r}$. Then

$$y'(x) = \sum_{n=0}^{\infty}(n + r)a_n x^{n+r-1} \quad \text{and} \quad y'' = \sum_{n=0}^{\infty}(n + r)(n + r - 1)a_n x^{n+r-2}.$$

Substituting these relations into the given ODE, we obtain

$$0 = \sum_{n=0}^{\infty}(n + r)(n + r - 1)a_n x^{n+r-1} - \sum_{n=0}^{\infty}(n + r)a_n x^{n+r}$$

$$- \sum_{n=0}^{\infty} 2(n + r)a_n x^{n+r-1} + \sum_{n=0}^{\infty} a_n x^{n+r}.$$

These summations can be written as

$$0 = \sum_{n=0}^{\infty}(n + r)(n + r - 3)a_n x^{n+r-1} - \sum_{n=0}^{\infty}(n + r - 1)a_n x^{n+r}.$$

Now we replace n by $(n - 1)$ in the second series and write the first term of the first series separately. We get

$$0 = r(r - 3)a_0 x^{r-1}$$

$$+ \sum_{n=1}^{\infty}\left((n + r)(n + r - 3)a_n - (n + r - 2)a_{n-1}\right)x^{n+r-1}.$$

We have

$$r(r - 3) = 0, \quad \text{i.e.} \quad r_1 = 0, \quad r_2 = 3.$$

Moreover,

$$(n + r)(n + r - 3)a_n - (n + r - 2)a_{n-1} = 0.$$

Thus,

$$a_n = \frac{(n + r - 2)a_{n-1}}{(n + r)(n + r - 3)}.$$

For $r = 3$, it is not difficult to show that

$$y_1(x) = 6a_0 \sum_{n=0}^{\infty} \frac{(n+1)x^{n+3}}{(n+3)!}.$$

For $r = 0$, we have

$$a_n = \frac{(n-2)a_{n-1}}{n(n-3)}, \quad n \geq 1,$$

that for $n = 1$ we obtain $a_1 = a_0/2$. We have $a_2 = 0$, but for $n = 3$, we obtain $a_3 = 0/0$. Without loss of generality, we assume $a_3 = c \equiv$ const and try to find the rest of the a_k in terms of c. Thus, we obtain

$$a_n = \frac{(2)(3)(4)\cdots(n-2)a_{n-1}}{((4)(5)\cdots(n))((1)(2)\cdots(n-3))} = \frac{6(n-2)c}{n!}, \quad n \geq 4.$$

Therefore,

$$y_2(x) = \sum_{n=0}^{\infty} a_n x^n = a_0 + a_1 x + a_2 x^2 + a_3 x^3 + \sum_{n=4}^{\infty} a_n x^n$$

$$= a_0 + \frac{a_0}{2}x + cx^3 + \sum_{n=4}^{\infty} \frac{6(n-2)c}{n!}x^n$$

$$= a_0 + \frac{a_0}{2}x + 6c \sum_{n=3}^{\infty} \frac{(n-2)x^n}{n!}$$

$$= a_0\left(1 + \frac{x}{2}\right) + 6c \sum_{n=3}^{\infty} \frac{(n-2)x^n}{n!}.$$

Note that if we replace n by $n + 3$ in the series, we obtain

$$y_2(x) = a_0\left(1 + \frac{x}{2}\right) + 6c \sum_{n=0}^{\infty} \frac{(n+1)x^n}{(n+3)!}.$$

It can be seen that the first solution $y_1(x)$ appears in the second solution. That is, if we start with a small value of r, we obtain the general solution.

$\square$

Example 6.17. Find two series solutions for the ODE

$$xy'' + xy' + y = 0.$$

Solution. Writing

$$y(x,r) = \sum_{n=0}^{\infty} a_n x^{n+r},$$

we find that

$$r(r-1)a_0 x^{r-1} + \big((n+r)(n+r-1)a_n + (n+r)a_{n-1}\big)x^{n+r-1} = 0, \quad n \ge 1.$$

The indicial roots are $r_1 = 0$ and $r_2 = 1$. We also have

$$a_n = -\frac{a_{n-1}}{n+r-1}, \quad n \ge 1.$$

Thus,

$$a_1 = -\frac{a_0}{r}, \quad a_2 = \frac{a_0}{r(r+1)}$$

and in general,

$$a_n = \frac{a_0 (-1)^n}{r(r+1)\cdots(n+r-1)}, \quad n \ge 2.$$

For $r = 1$, it is not difficult to show that

$$y_1(x) = a_0 x \sum_{n=0}^{\infty} \frac{(-1)^n x^n}{n!} = a_0 x e^{-x}.$$

On the other hand, for $r = 0$ we obtain $0 \cdot a_1 = -a_0$, i.e., $a_0 = 0$. But we assumed $a_0 \ne 0$. To avoid this difficulty, we proceed as we have done for Case 2. We assume that the second solution $y_2(x)$ is logarithmic. Since in Case 2 we had a multiple root, we choose $a_0 = r$ which implies that $a_1 = -a_0/r = -1$ and

$$a_n = \frac{a_0 (-1)^n}{r(r+1)\cdots(n+r-1)} = \frac{(-1)^n}{(r+1)\cdots(n+r-1)}, \quad n \ge 2.$$

We have

$$\ln(a_n) = \ln\left((-1)^n\right) - \left(\ln(r+1) + \ln(r+2) + \cdots + \ln(n+r-1)\right), \quad n \geq 2.$$

Differentiating with respect to r gives

$$a_n' = -a_n \left(\frac{1}{1+r} + \frac{1}{2+r} + \cdots + \frac{1}{n+r-1} \right), \quad n \geq 2.$$

If we substitute $r = 0$, we obtain

$$a_n' = \frac{(-1)^{n-1} H_{n-1}}{(n-1)!}, \quad n \geq 2,$$

where

$$H_{n-1} \equiv \frac{1}{1} + \frac{1}{2} + \cdots + \frac{1}{n-1}.$$

Therefore, the second linearly independent solution will be

$$y_2(x) = y_1(x)\ln(x) + x^{-1} - 1 + \sum_{n=2}^{\infty} \frac{(-1)^{n-1} H_{n-1}}{(n-1)!} x^{n-1}.$$

□

Remark 6.18. A series solution may not be of much practical importance for a large value of x since too many terms in the series may be required to obtain a desired degree of accuracy. In this case, we make the substitutions

$$t = \frac{1}{x}, \quad \frac{dy}{dx} = -t^2 \frac{dy}{dt}, \quad \frac{d^2 y}{dx^2} = t^4 \frac{d^2 y}{dt^2} + 2t^3 \frac{dy}{dt},$$

and solve the equation in a neighborhood of $t = 0$. □

Example 6.19. Using the substitutions given in Remark 6.18, find the general solution of the ODE

$$2x^3 y'' - x^2 y' + 3y = 0$$

for large values of x.

Solution. By using the above substitutions, we can represent the ODE in terms of t. We obtain

$$2t \frac{d^2 y}{dt^2} + 5 \frac{dy}{dt} + 3y = 0.$$

The point $t = 0$ is a regular singular point. Therefore, we set

$$y(t) = \sum_{n=0}^{\infty} a_n t^{n+r},$$

and we have

$$y'(t) = \sum_{n=0}^{\infty} (n + r) a_n t^{n+r-1} \text{ and } y''(t) = \sum_{n=0}^{\infty} (n + r)(n + r - 1) a_n t^{n+r-2}.$$

Substituting these expressions into the transformed ODE leads to

$$0 = 2 \sum_{n=0}^{\infty} (n + r)(n + r - 1) a_n t^{n+r-1} + 5 \sum_{n=0}^{\infty} (n + r) a_n t^{n+r-1}$$

$$+ 3 \sum_{n=0}^{\infty} a_n t^{n+r} = 0.$$

We write the first and the second series with one symbol and obtain

$$0 = \sum_{n=0}^{\infty} (n + r)(2n + 2r + 5) a_n t^{n+r-1} + 3 \sum_{n=0}^{\infty} a_n t^{n+r}.$$

Re-indexing the first series by 1 and writing separately the term for $n = 0$ of the second series, we get

$$0 = r(2r + 5) a_0 t^{r-1} + \sum_{n=1}^{\infty} (n + r)(2n + 2r + 5) a_{n-1} t^{n+r-1}$$

$$+ 3 \sum_{n=0}^{\infty} a_n t^{n+r-1}.$$

Writing these two series with one symbol, we obtain

$$0 = r(2r + 5) a_0 t^{r-1} + \sum_{n=1}^{\infty} \left((n + r)(2n + 2r + 5) a_{n-1} + 3 a_n \right) t^{n+r-1}.$$

Since $a_0 \neq 0$, it holds $r(2r + 5) = 0$, i.e., $r_1 = 0$ and $r_2 = -5/2$.
    We also have

$$(n + r)(2n + 2r + 5) a_{n-1} + 3 a_n = 0.$$

That is,

$$a_n = \frac{-3a_{n-1}}{(n+r)(2n+2r+5)}, \quad n \geq 1.$$

The roots do not differ by an integer, so we have the Case 2. We start with $r_1 = 0$ and obtain

$$a_n = -\frac{3a_{n-1}}{n(2n+5)} = \frac{120(-3)^n a_0}{n!(7 \cdot 9 \cdot 11 \cdots (n+5))}, \quad n \geq 1.$$

Therefore,

$$y_1(x) = \sum_{n=0}^{\infty} a_n x^n = a_0 + \sum_{n=1}^{\infty} a_n x^n = a_0 + \sum_{n=1}^{\infty} \frac{(-3)^n a_0 x^n}{n!(7 \cdot 9 \cdot 11 \cdots (n+5))}$$

$$= a_0 \left(1 + \sum_{n=1}^{\infty} \frac{(-3)^n x^n}{n!(7 \cdot 9 \cdot 11 \cdots (n+5))} \right).$$

In a similar manner, for $r_2 = -5/2$, we obtain

$$y_2(x) = a_0 x^{-\frac{5}{2}} \left(1 + \sum_{n=1}^{\infty} \frac{(-3)^n x^n}{n!((-3) \cdot (-1) \cdot (1) \cdots (2n-5))} \right).$$

$\square$

6.5   Special Functions

In advanced work in engineering, physics, and applied mathematics, a few special linear second-order ODEs occur which are of particular interest. There is an extensive literature dealing with the theory and application of these equations. Three of these ODEs are Legendre's equation, Bessel's equation, and the hypergeometric equation. The solutions of these and other equations that occur in applications are often referred to as *special functions*. We shall briefly consider the three abovementioned equations.

6.5.1   Legendre's Equation

The general form of Legendre's equation is given by

$$(1 - x^2)y'' - 2xy' + \lambda(\lambda + 1)y = 0,$$

where λ is a real constant. Obviously, $x = 0$ is an ordinary point for this equation. Therefore, it can be solved by the power series method.

Let

$$y(x) = \sum_{n=0}^{\infty} a_n x^n.$$

By forming $y'(x)$ and $y''(x)$ and substituting the resulting expressions into the given equation, we obtain

$$0 = (1 - x^2) \sum_{n=0}^{\infty} n(n-1)a_n x^{n-2} - 2x \sum_{n=0}^{\infty} na_n x^{n-1} + \lambda(\lambda + 1) \sum_{n=0}^{\infty} a_n x^n.$$

Simplifying this equation gives

$$0 = \sum_{n=0}^{\infty} n(n-1)a_n x^{n-2} - \sum_{n=0}^{\infty} (n - \lambda)(n + \lambda + 1)a_n x^n.$$

Shifting the index in the second series by lowering the index range by 2 and writing the first two terms of the first series separately, we obtain

$$0 = 0 + 0 + \sum_{n=2}^{\infty} n(n-1)a_n x^{n-2} - \sum_{n=2}^{\infty} (n - \lambda - 2)(n + \lambda - 1)a_{n-2} x^{n-2}.$$

We write these two series with one symbol and obtain

$$0 = \sum_{n=2}^{\infty} \left(n(n-1)a_n - (n - \lambda - 2)(n + \lambda - 1)a_{n-2} \right) x^{n-2}.$$

Since $x^{n-2} \neq 0$, we have

$$n(n-1)a_n - (n - \lambda - 2)(n + \lambda - 1)a_{n-2} = 0, \quad n \geq 2,$$

i.e.,

$$a_n = -\frac{(\lambda - n + 2)(n + \lambda - 1)a_{n-2}}{n(n-1)}, \quad n \geq 2. \tag{6.16}$$

If we choose $n = 2, 3, 4, \ldots$, we obtain

$$a_2 = -\frac{\lambda(\lambda + 1)}{2!}a_0,$$

$$a_3 = -\frac{(\lambda - 1)(\lambda + 2)}{3!}a_1,$$

$$a_4 = \frac{(\lambda - 2)\lambda(\lambda + 1)(\lambda + 3)}{4!}a_0,$$

$$a_5 = \frac{(\lambda - 3)(\lambda - 1)(\lambda + 2)(\lambda + 4)}{5!}a_1,$$

$$a_6 = -\frac{(\lambda - 4)(\lambda - 2)\lambda(\lambda + 1)(\lambda + 3)(\lambda + 5)}{6!}a_0,$$

$$a_7 = -\frac{(\lambda - 5)(\lambda - 3)(\lambda - 1)(\lambda + 2)(\lambda + 4)(\lambda + 6)}{7!}a_1,$$

$$\vdots$$

Therefore, we have

$$y(x) = \sum_{n=0}^{\infty} a_n x^n = a_0 + a_1 x + a_2 x^2 + a_3 x^3 + a_4 x^4 + \cdots$$

$$= a_0 \left(1 - \frac{\lambda(\lambda + 1)}{2!}x^2 + \frac{(\lambda - 2)\lambda(\lambda + 1)(\lambda + 3)}{4!}x^4 - \cdots \right)$$

$$+ a_1 \left(x - \frac{(\lambda - 1)(\lambda + 2)}{3!}x^3 + \frac{(\lambda - 3)(\lambda - 1)(\lambda + 2)(\lambda + 4)}{5!} - \cdots \right).$$

$$+ \cdots$$

This solution is valid for $|x| < 1$. Let $a_0 = a_1 = 1$. Then we obtain

$$y_1(x) = 1 - \frac{\lambda(\lambda + 1)}{2!}x^2 + \frac{(\lambda - 2)\lambda(\lambda + 1)(\lambda + 3)}{4!}x^4 - \cdots$$

$$y_2(x) = x - \frac{(\lambda - 1)(\lambda + 2)}{3!}x^3 + \frac{(\lambda - 3)(\lambda - 1)(\lambda + 2)(\lambda + 4)}{5!} - \cdots.$$

When λ is zero or an integer, then the two series terminate, i.e., they become polynomials in x. For example, for $\lambda = 0, \ldots, 5$, we obtain

$$\lambda = 0: \quad y_1(x) = 1,$$

$$\lambda = 1: \quad y_2(x) = x,$$

$$\lambda = 2: \quad y_1(x) = 1 - 3x^2,$$

$$\lambda = 3: \quad y_2(x) = x - \frac{5}{3}x,$$

$$\lambda = 4: \quad y_1(x) = 1 - 10x^2 + \frac{35}{3}x^4,$$

$$\lambda = 5: \quad y_2(x) = x - \frac{14}{3}x + \frac{21}{5}x^5.$$

Since $cy_1(x)$ or $cy_2(x)$, $c \in \mathbb{R}$, are also solutions of the (homogeneous) Legendre's equation, we choose c such that for $x = 1$ all polynomials become one. We denote these polynomials by $P_n(x)$ which are known as Legendre polynomials. Thus, we have

$$P_0(x) = 1, \quad P_1(x) = x, \quad P_2(x) = \frac{1}{2}(3x^2 - 1), \quad P_3(x) = \frac{1}{2}(5x^3 - 3x),$$

$$P_4(x) = \frac{1}{8}(35x^4 - 30x^2 + 3), \quad P_5(x) = \frac{1}{8}(63x^5 - 70x^3 + 15x).$$

If we write the recurrence relation (6.16) in the form

$$a_{n+2} = -\frac{(\lambda - n)(n + \lambda + 1)}{(n + 1)(n + 2)}a_n, \quad n \geq 0, \tag{6.17}$$

then for $\lambda = n$, we obtain $a_{n+2} = a_{n+4} = a_{n+6} = \cdots = 0$. Hence, one of the two solutions will be a polynomial with degree n and the other one is a series solution. Using (6.17), we can conclude that

$$P_n(x) = \sum_{k=0}^{n}(-1)^k \frac{(2\lambda - 2k)x^{\lambda-2k}}{2^\lambda k!(\lambda - k)!(\lambda - 2k)!}, \tag{6.18}$$

where $n = \lambda/2$ if λ is even and $n = (\lambda - 1)/2$ if λ is odd. A compact formula for expressing the Legendre polynomials is

$$P_n(x) = \frac{1}{2^n n!} \frac{d^n}{dx^n}(x^2 - 1)^n, \quad n \geq 0. \tag{6.19}$$

This formula is known as *Rodrigues' formula*.

An important property of the Legendre polynomials is its orthogonality, i.e., it holds

$$\int_{-1}^{1} P_m(x) P_n(x)\, dx = 0, \quad m \neq n.$$

For $m = n$, we have

$$\int_{-1}^{1} (P_n(x))^2 dx = \frac{2}{2n + 1}, \quad n \geq 0.$$

Finally, it is not difficult to show that the Legendre polynomials satisfy the recurrence formula

$$(n + 1) P_{n+1}(x) = (2n + 1)x P_n(x) - n P_{n-1}(x), \quad n \geq 1.$$

There are more properties for Legendre polynomials. We leave them as exercises at the end of this chapter.

6.5.2   Bessel's Equation

The ODE

$$x^2 y'' + xy' + (x^2 - \lambda^2)y = 0, \tag{6.20}$$

where λ is a nonnegative real constant, is known as Bessel's equation. It has a regular singular point at $x = 0$. Hence, a series solution for (6.20) can be obtained by the method of power series for $0 < x < \infty$. We shall seek the series solutions in the form

$$y(x) = \sum_{n=0}^{\infty} a_n x^{n+r}.$$

Some computation leads to

$$0 = (r^2 - \lambda^2) a_0 x^r + ((r + 1)^2 - \lambda^2) a_1 x^{r-1}$$

$$+ \sum_{n=2}^{\infty} (((r + n)^2 - \lambda^2) a_n + a_{n-2}) x^{n+r}.$$

We have $r_1 = \lambda,\ r_2 = -\lambda$ and

$$a_n = -\frac{a_{n-2}}{(r+n)^2 - \lambda^2}, \quad n \geq 2. \tag{6.21}$$

The case $r = \lambda$ implies $a_1 = 0$ and

$$a_n = -\frac{a_{n-2}}{n(n+2\lambda)}, \quad n \geq 2.$$

Thus, we obtain $a_3 = a_5 = a_7 = \cdots = 0$. For even values of n, we have

$$a_{2n} = \frac{(-1)^n a_0}{2^{2n} n! (\lambda+1)(\lambda+2)\cdots(\lambda+n)}, \quad n \geq 0.$$

Since a_0 is arbitrary, we take

$$a_0 = \frac{1}{2^\lambda \Gamma(1+\lambda)}$$

to satisfy the orthogonality condition. Then it follows that

$$a_{2n} = \frac{(-1)^n}{2^{2n+\lambda} n! \Gamma(1+\lambda+n)}, \quad n \geq 0, \tag{6.22}$$

where

$$\Gamma(1+\lambda+n) = (\lambda+1)(\lambda+2)\cdots(\lambda+n)\Gamma(1+\lambda).$$

Substituting (6.22) into

$$y(x) = \sum_{n=0}^{\infty} a_n x^{n+r}$$

(where n is replaced by $2n$), we obtain a solution. We denote it by $J_\lambda(x)$. Therefore,

$$J_\lambda(x) = \sum_{n=0}^{\infty} \frac{(-1)^n \left(\frac{x}{2}\right)^{2n+\lambda}}{n! \Gamma(1+\lambda+n)}. \tag{6.23}$$

We call $J_\lambda(x)$ the *Bessel function of the first kind of index* λ. From this formula, we observe that $J_0(0) = 1$ and $J_{-\lambda}(0) = 0$ for $\lambda > 0$.

Similarly, in the case $r = -\lambda$, we take

$$a_0 = \frac{1}{2^{-\lambda} \Gamma(1-\lambda)}$$

and obtain a second solution

$$J_{-\lambda}(x) = \sum_{n=0}^{\infty} \frac{(-1)^n \left(\frac{x}{2}\right)^{2n-\lambda}}{n!\Gamma(1-\lambda+n)}. \tag{6.24}$$

If λ is not an integer, then (6.23) and (6.24) are linearly independent. Thus, the general solution of (6.21) is $y(x) = AJ_{\lambda}(x) + BJ_{-\lambda}(x)$.

The Bessel function of the first kind satisfies several useful recurrence formulae. Examples include

1. $\dfrac{d}{dx}\left(x^{\lambda+1}J_{\lambda+1}(x)\right) = x^{\lambda+1}J_{\lambda}(x).$

2. $\dfrac{d}{dx}\left(x^{-\lambda}J_{\lambda}(x)\right) = -x^{-\lambda}J_{\lambda+1}(x).$

3. $\dfrac{dJ_{\lambda}}{dx}(x) + \dfrac{\lambda}{x}J_{\lambda}(x) = J_{\lambda-1}(x).$

4. $\dfrac{dJ_{\lambda}}{dx}(x) - \dfrac{\lambda}{x}J_{\lambda}(x) = -J_{\lambda+1}(x).$

5. $J_{\lambda+1}(x) = \dfrac{2\lambda}{x}J_{\lambda}(x) - J_{\lambda-1}(x).$

6. $J_{\lambda-1}(x) = \dfrac{2\lambda}{x}J_{\lambda}(x) - J_{\lambda+1}(x).$

7. $J_{\lambda}'(x) = \dfrac{1}{2}\left(J_{\lambda+1}(x) - J_{\lambda-1}(x)\right).$

If λ is a nonnegative integer, then $J_{\lambda}(x)$ is still a solution, but $J_{-\lambda}(x)$ is not, because it can be seen that

$$\frac{1}{\Gamma(\lambda)} = 0 \quad \text{for } \lambda = 0, -1, -2, \dots.$$

Hence, (6.24) becomes

$$J_{-\lambda}(x) = \sum_{n=0}^{\infty} \frac{(-1)^n \left(\frac{x}{2}\right)^{2n-\lambda}}{n!\Gamma(1-\lambda+n)} = \sum_{n=\lambda}^{\infty} \frac{(-1)^n \left(\frac{x}{2}\right)^{2n-\lambda}}{n!\Gamma(1-\lambda+n)}.$$

By replacing $n - \lambda$ by n, it is clear that $J_{-\lambda}(x) = (-1)^{\lambda}J_{\lambda}(x)$, i.e., $J_{\lambda}(x)$ and $J_{-\lambda}(x)$ are linearly dependent.

In general if $\lambda = N$ is an integer number, a second solution that we denote by $Y_N(x)$ can be found by the following formula

$$Y_N(x) = \frac{2}{\pi} J_N(x) \left(\ln\left(\frac{x}{2}\right) + \gamma \right) - \frac{1}{\pi} \sum_{n=2}^{\infty} \frac{(-1)^n \left(\frac{x}{2}\right)^{2n+N}}{n!(N+n)!} (H_{n+N} + H_n)$$

$$- \frac{1}{\pi} \sum_{n=0}^{\infty} \frac{(N-n-1)^n (x)^{2n-N}}{2^{2n+N} n!},$$

where $\gamma = 0.5772156649\ldots$ is known as the *Euler's constant* and as before

$$H_n \equiv 1 + \frac{1}{2} + \frac{1}{3} + \cdots + \frac{1}{n}.$$

The solution $Y_N(x)$ is called *Bessel function of the second kind of index* λ.

6.5.3 Hypergeometric Equation

Let the linear second-order ODE

$$x(1-x)y''(x) + \big(c - (a+b+1)x\big)y'(x) - aby(x) = 0 \qquad (6.25)$$

be given, where a, b, and c are fixed real parameters. This equation has three singular points (one may be at infinity) which are regular. The indicial roots relative to $x = 0$ are $r_1 = 0$ and $r_2 = 1 - c$; those relative to $x = 1$ are $r_1 = 0$ and $r_2 = c - a - b$.

For $x = 0$ with $r_1 = 0$, we get

$$a_n = \frac{(n+a-1)(n+b-1)a_{n-1}}{n(n+c-1)}, \qquad n \geq 1.$$

Solving recursively for a_n, we obtain

$$a_n = \frac{a(a+1)\cdots(n+a-1)b(b+1)\cdots(n+b-1)a_0}{n!c(c+1)\cdots(n+c-1)}, \qquad n \geq 1.$$

If we use the *factorial function* $(a)_n$ which is defined for nonnegative n by

$$(a)_n \equiv a(a+1)\cdots(n+a-1), \qquad n \geq 1,$$

and $(a)_0 \equiv 1$ when $a \neq 0$, then we can write

$$a_n = \frac{(a)_n (b)_n a_0}{n!(c)_n}, \qquad n \geq 1.$$

In terms of factorial functions, the first solution of the hypergeometric equation reads

$$y_1(x) = 1 + \sum_{n=1}^{\infty} \frac{(a)_n (b)_n x^n}{n!(c)_n}. \tag{6.26}$$

This solution is called *hypergeometric function* and denoted by $F(a, b, c; x)$.

Similarly, if we study the case $r = 1 - c$ and seek a solution in the form

$$y(x) = \sum_{n=0}^{\infty} b_n x^{n+1-c},$$

we obtain

$$y_2(x) = x^{1-c} + \sum_{n=1}^{\infty} \frac{(a+1-c)_n (b+1-c)_n x^{n+1-c}}{n!(2-c)_n}. \tag{6.27}$$

One may show that

$$y_2(x) = x^{1-c} F(a+1-c, b+1-c, 2-c; x).$$

Note that for $y_1(x)$ and $y_2(x)$, we have set $a_0 \equiv 1$ and $b_0 \equiv 1$, respectively.

When c is an integer (corresponding to the larger root, 0 or $1 - c$), one of the solutions can be (6.26) or (6.27). The second solution may or may not have a logarithmic term. If a or b is zero or a negative integer, the hypergeometric series terminates and the hypergeometric function is a polynomial.

We list, without proof, a number of properties of the hypergeometric functions:

1. $F(a, b, c; x) = F(b, a, c; x)$.
2. $F(n, b, b; x) = (1 - x)^{-n}$.
3. $F(1, 1, 2; -x) = x^{-1} \ln(1 + x)$.
4. $F\left(\frac{1}{2}, \frac{1}{2}, \frac{3}{2}; x^2\right) = x^{-1} \sin^{-1}(x)$.
5. $F\left(\frac{1}{2}, 1, \frac{3}{2}; -x^2\right) = x^{-1} \tan^{-1}(x)$.

Remark 6.20. In general, if $x = 1 - t$, then (6.25) becomes

$$t(1-t)\frac{d^2 y}{dt^2}(t) + \left(a + b - c + 1 - (a + b + 1)t\right)\frac{dy}{dt}(t) - ab\, y(t) = 0,$$

where its general solution is

$$y(t) = AF(a, b, a + b - c + 1; 1 - t)$$

$$+ B(1 - t)^{c-a-b} F(c - b, c - a, c - a - b + 1; 1 - t).$$

□

Remark 6.21. Sometimes it will be possible to transform a differential equation into a hypergeometric equation as is shown in the next example. □

Example 6.22. Find the general solution about $x = 0$ of the ODE

$$(e^{-x} - 1)y''(x) + y'(x) - e^{-x}y(x) = 0.$$

Solution. Let $e^{-x} = t$, then

$$\frac{dy}{dx} = -t\frac{dy}{dt} \quad \text{and} \quad \frac{d^2y}{dx^2} = t^2\frac{d^2y}{dt^2} + t\frac{dy}{dt}.$$

Substituting these expressions into the given ODE leads to

$$t(1 - t)\frac{d^2y}{dt^2}(t) + (2 - t)y'(t) + y(t) = 0.$$

This is a hypergeometric equation with $a = 1$, $b = -1$, and $c = 3$. Since for $x = 0$ we have $t = 1$, the general solution at $t = 1$ is

$$y(t) = AF(1, -1, -2; 1 - t) + B(1 - t)^3 F(4, 2, 4; 1 - t).$$

That is,

$$y(t) = A\left(1 + \sum_{n=1}^{\infty} \frac{(1)_n(-1)_n(1 - t)^n}{n!(3)_n}\right)$$

$$+ B\left((1 - t)^{-1} + \sum_{n=1}^{\infty} \frac{(-1)_n(-3)_n(1 - t)^n}{n!(-1)_n}\right).$$

Now, we replace t by e^{-x} and obtain

$$y(x) = A\left(1 + \sum_{n=1}^{\infty} \frac{(1)_n(-1)_n(1 - e^{-x})^n}{n!(3)_n}\right)$$

$$+ B\left((1 - e^{-x})^{-1} + \sum_{n=1}^{\infty} \frac{(-1)_n(-3)_n(1 - e^{-x})^n}{n!(-1)_n}\right).$$

□

6.6   Exercises

Exercise 6.1. For the following ODEs, classify each singular point (real or complex) as regular or irregular.

(1) $x(1 - x^2)y'' + \sin(x)y' + y = 0$.   (2) $(1 + x^2)y'' + xy' - 3y = 0$.

(3) $x^3 y'' + x^2 y' + 4y = 0$.   (4) $x^2(1 - x)y'' + y' - xy = 0$.

(5) $x^2(1 - x)y'' + y' - xy = 0$.   (6) $x^3(x + 1)y'' + xy' + y = 0$.

(7) $(x^2 - 4)^2 y'' + y' + 5x^2 y = 0$.   (8) $x^3 y'' + x(x - 1)y' - 6y = 0$.

(9) $x^3 y'' - x\sin(x)y' + \tan(x)y = 0$.

(10) $x^4 y'' + 2x^2 y' - (2x - 1)y = 0$.

Exercise 6.2. Find the series solution for the following ODEs by the method of undetermined coefficients.

(1) $y' - x^3 = e^x$.   (2) $(1 - xy)y' + y = e^x$.

(3) $y'' + \sin(x)y' + e^x y = 0$.   (4) $(1 + x^2)y'' - xy' + y = \sin(x)$.

(5) $y'' - xy' - y = \sinh(x)$.   (6) $x^2 y'' + \sin(x)y' - \cos(x)y = 0$.

(7) $y'' - xy' + e^x y = x$.   (8) $y' + xy + x^2 = 0$.

(9) $x^2 y'' = 2x - 1$.   (10) $y'' - 4y' + 4y = 3e^{2x}$.

Exercise 6.3. Let the point $x = 0$ be an ordinary point. Express the general solution in terms of a power series about this point.

(1) $(1 + 5x^2)y'' - 10y = 0$.   (2) $y'' - 2xy' + 4y = 0$.

(3) $(1 - x^2)y'' - 2xy' + 12y = 0$.   (4) $(4 + x^2)y'' + 2xy' - 4xy = 0$.

(5) $y'' - 2xy' + 6y = 0$.   (6) $y'' - 2xy = 0$.

(7) $(1 + 2x^2)y'' + 11xy' + 9y = 0$.   (8) $2y'' - 5xy' + 2y = 0$.

(9) $2y'' + 9xy' - 36y = 0$.   (10) $(x^2 + 1)y'' + 3xy' - 3y = 0$.

Exercise 6.4. Use the method discussed in Sect. 6.4 to obtain the series solution about $x = 0$ for two linearly independent solutions.

- Case 1

  (1) $3xy'' + 2y' - 4y = 0$.   (2) $2xy'' - y' + y = 0$.

  (3) $2xy'' + (3 + x)y' - y = 0$.   (4) $2xy'' + (1 - 2x)y' + 2y = 0$.

  (5) $2x^2 y'' - 3x(1 - x)y' + y = 0$.   (6) $2xy'' + (x + 1)y' + 3y = 0$.

  (7) $3x^2 y'' + 2xy' - (2 + x)y = 0$.   (8) $4xy'' + y' + y = 0$.

  (9) $6xy'' - 2y' + y = 0$.   (10) $5y'' + 3xy' - 4y = 0$.

- Case 2

(11) $x^2(1-x)y'' + 3xy' - 4y = 0.$ (12) $xy'' + y' - 2y = 0.$

(13) $xy'' + (1-x^2)y' - xy = 0.$ (14) $xy'' + y' + xy = 0.$

(15) $xy'' + (1+x)y' + 2y = 0.$ (16) $xy'' + y' + x^2y = 0.$

(17) $x^2y'' + (1-2x)y' + y = 0.$ (18) $4x^2y'' + (1+3x)y = 0.$

(19) $x^2y'' + x(3+2x)y' + y = 0.$

(20) $x^2y'' + 3xy' + (1+4x^2)y = 0.$

- Case 3

(21) $xy'' - (4+x)y' + 2y = 0.$ (22) $x^2y'' + x^2y' - 2y = 0.$

(23) $xy'' - 2(2+x)y' + 4y = 0.$ (24) $xy'' + y = 0.$

(25) $x^2y'' + xy' + (x^2-1)y = 0.$ (26) $xy'' + xy' + y = 0.$

(27) $xy'' - y' + y = 0.$ (28) $xy'' - xy' - y = 0.$

(29) $x^2y'' - x(6+x)y' + 10y = 0.$

(30) $x^2y'' + xy' + \left(x^2 - \dfrac{1}{9}\right)y = 0.$

Exercise 6.5. Verify that the point at infinity is either an ordinary point or singular point. Express the general solution in the form of a series of powers of $1/x$.

(1) $2x^3y'' - x(2-5x)y' + y = 0.$ (2) $x(1-x)y'' - 3y' + 2y = 0.$

(3) $x^4y'' + x^2y' + y = 0.$ (4) $2x^3y'' - x(2-5x)y' + y = 0.$

(5) $xy'' - y' + y = 0.$ (6) $x^4y'' + x\left(1+2x^2\right)y' + 5y = 0.$

Exercise 6.6. Express a general solution to the given hypergeometric equations:

(1) $x(x-1)y'' - \left(3x - \dfrac{1}{2}\right)y' + y = 0.$

(2) $2x(x-1)y'' + (3-x)y' + 2y = 0.$

(3) $x(1-x)y'' + \left(\dfrac{3}{2} - 2x\right)y' - \dfrac{1}{4}y = 0.$

(4) $x(1-x)y'' + 2(1-2x)y' - 2y = 0.$

Exercise 6.7. Prove that

(1) $F(1,1,2;-x) = x^{-1}\ln(1+x).$ (2) $2F(2,b,b;x) = (1-x)^{-1}.$

(3) $J_{\lambda+1}(x) + J_{\lambda-1}(x) = \dfrac{2\lambda}{x}J_\lambda(x).$ (4) $\dfrac{d}{dx}\left(x^\lambda J_\lambda(x)\right) = x^\lambda J_{\lambda-1}(x).$

(5) $F\left(\tfrac{1}{2}, 1, \tfrac{3}{2}; -x^2\right) = x^{-1}\tan^{-1}(x).$ (6) $J_{\frac{1}{2}}(x) = \sqrt{\dfrac{2}{\pi x}}\sin(x).$

(7) $F\left(\frac{1}{2}, \frac{1}{2}, \frac{3}{2}; x^2\right) = x^{-1}\sin^{-1}(x)$.   (8) $J_{\lambda+1}(x) - J_{\lambda-1}(x) = 2J'_\lambda(x)$.

(9) $\dfrac{dJ_\lambda}{dx} - \dfrac{\lambda}{x}J_\lambda = -J_{\lambda+1}$.   (10) $J_{-\frac{1}{2}}(x) = \sqrt{\dfrac{2}{\pi x}}\cos(x)$.

(11) $J_0(x) = \dfrac{1}{\pi}\displaystyle\int_0^\pi \cos(x\cos(\theta))d\theta$.   (12) $\displaystyle\int_0^\infty J_0(x)dx = 1$.

(13) $\mathcal{L}[J_0(x)] = \dfrac{1}{\sqrt{s^2+1}}$.   (14) $\mathcal{L}[J_1(x)] = 1 - \dfrac{s}{\sqrt{s^2+1}}$.

Exercise 6.8. The ODE

$$xy'' + (1-x)y' + \lambda y = 0, \tag{6.28}$$

where λ is a real constant, is known as Laguerre's equation. Show that in the case when λ is a nonnegative integer this equation has a solution that is a polynomial of degree λ given by

$$L_\lambda(x) \equiv \lambda!\sum_{n=0}^\lambda \frac{(-1)^n x^n}{(n!)^2(\lambda-n)!}.$$

Show that

$$\int_0^\infty e^{-x}L_m(x)L_n(x)dx = \begin{cases} 0, & m \neq n \\ 1, & m = n. \end{cases}$$

By induction prove that Rodrigues' formula for Laguerre's equation reads

$$L_\lambda(x) = \frac{e^x}{\lambda!}\frac{d^\lambda}{dx^\lambda}\left(x^\lambda e^{-x}\right).$$

Prove that

$$L_{n+1}(x) - (2n+1-x)L_n(x) + n^2 L_{n-1}(x) = 0.$$

Exercise 6.9. The ODE

$$y'' - 2xy' + 2\lambda y = 0, \tag{6.29}$$

where λ is a nonnegative integer, is known as Hermite's equation. Show that this equation has a solution that is a polynomial of degree λ given by

$$H_\lambda(x) \equiv \lambda!\sum_{n=0}^\lambda \frac{(-1)^n (2x)^{\lambda-2n}}{n!(\lambda-2n)!}.$$

Prove that

(1) $-H_n(-x) = (-1)^n H_n(x)$.   (2) $-H_{2n+1}(0) = 0$.

(3) $-e^{2xt-t^2} = \sum_{n=0}^{\infty} \frac{H_n(x)}{n!} t^n$.   (4) $-L_{2n} = (-1)^n e^{x^2} \frac{d^n}{dx^n}\left(e^{-x^2}\right)$.

(5) $-H_{2n}(0) = (-1)^n 2^{2n}(1 \cdot 3 \cdot 5 \cdots (2n-1))$.

(6) $-\int_{-\infty}^{\infty} e^{-x^2} H_m(x) H_n(x) dx = \begin{cases} 0, & m \neq n \\ 2^n n! \sqrt{\pi}, & m = n \end{cases}$.

Exercise 6.10. Consider the ODE

$$(1 - x^2)y'' - xy' + \lambda^2 y = 0, \tag{6.30}$$

which is known as Chebyshev's equation that is important in approximation theory. It has an ordinary point at $x = 0$. For $\lambda = n$, it possesses a polynomial solution $T_n(x)$ which satisfies

$$T_n(x) = \cos(n \arccos(x)).$$

It is clear that $T_0(x) = 1$ and $T_1(x) = x$. Show that

(1) $-T_{n+m}(x) + T_{n-m}(x) = 2T_n(x)T_m(x)$.
(2) $-T_{n+1}(x) + T_{n-1}(x) = 2xT_n(x)$.

Prove that

$$\left(\sqrt{1-x^2}\, T_n'(x)\right)' + \frac{n^2}{\sqrt{1-x^2}} T_n(x) = 0.$$

Use this result and show that

$$\int_{-1}^{1} \frac{T_m(x)T_n(x)}{\sqrt{1-x^2}} dx = \begin{cases} 0, & m \neq n \\ \dfrac{\pi}{2}, & m = n \neq 0 \\ \pi, & m = n = 0. \end{cases}$$

Prove that

$$T_n(x) = \frac{(-2)^n n! (1-x^2)^{\frac{1}{2}}}{(2n)!} \frac{d^n}{dx^n}(1-x^2)^{n-\frac{1}{2}}.$$

Exercise 6.11. The ODE

$$y'' + xy = 0 \tag{6.31}$$

is known as Airy's equation. It has a power series expansion about $x = 0$. Find the general solution of this equation. Moreover, show that

(i) The substitution $y = u\sqrt{x}$ changes this equation into

$$x^2 u'' + xu' + \left(x^3 - \frac{1}{4}\right)u = 0.$$

(ii) The substitution $x = \sqrt[3]{\frac{9}{4}r^2}$ changes the equation in part (i) into

$$r^2 u'' + ru' + \left(r^2 - \frac{1}{9}\right)u = 0$$

which is a Bessel's equation with $\lambda = 1/3$.

Use parts (i) and (ii) to show that

$$y(x) = c_1 \sqrt{x} J_{\frac{1}{3}}\left(\frac{2}{3}x\sqrt{x}\right) + c_1 \sqrt{x} J_{-\frac{1}{3}}\left(\frac{2}{3}x\sqrt{x}\right).$$

Exercise 6.12. Show that the ODE

$$x^4 y'' + 2x^2 y' - (2x - 1)y = 0$$

has an irregular singular point at $x = 0$. Therefore, there does not exist a series solution for this ODE. Solve it by the generalized method of order reduction discussed in Chap. 3.

Exercise 6.13. Solve the following problems:

- Consider Bessel's equation with index zero. Suppose $y_1(x) = J_0(x)$. Use the method of order reduction to find the second solution.
- If we have Bessel's equation with index one and $y_1(x) = J_1(x)$, determine the second solution by the method of order reduction.

Reference

1. Boyce, W.E., DiPrima, R.C.: Elementary Differential Equations. Wiley, Chichester et al. (2012)

Chapter 7
Numerical Methods for Initial Value Problems

7.1 Initial Value Problems

Let us consider the following system of n first-order ODEs:

$$y_1'(x) = f_1(x, y_1(x), y_2(x), \ldots, y_n(x)),$$
$$y_2'(x) = f_2(x, y_1(x), y_2(x), \ldots, y_n(x)),$$
$$\ldots$$
$$y_n'(x) = f_n(x, y_1(x), y_2(x), \ldots, y_n(x)), \qquad (7.1)$$

where $y_i'(x)$ denotes the derivative of the function $y_i(x)$ w.r.t. x, $i = 1, \ldots, n$.

Setting $y(x) \equiv (y_1(x), \ldots, y_n(x))^T$ and $f(x, y) \equiv (f_1(x, y), \ldots, f_n(x, y))^T$, this system can be formulated in vector notation as

$$y'(x) \equiv \frac{d}{dx} y(x) = f(x, y(x)). \qquad (7.2)$$

Assume that $x \in J \subset \mathbb{R}$ and $y(x) \in \Omega \subset \mathbb{R}^n$. The mapping $f : J \times \Omega \to \mathbb{R}^n$ is called *vector field* and $y(x)$ is the *solution* of the ODE.

In the previous chapters, we have encountered higher-order ODEs. Such a higher-order equation

$$u^{(n)}(x) = G\left(x, u(x), u'(x), u''(x), \ldots, u^{(n-1)}(x)\right) \qquad (7.3)$$

can be transformed into the system (7.1) by setting

$$y = (y_1, y_2, \ldots, y_n)^T \equiv \left(u, u', u'', \ldots, u^{(n-1)}\right)^T.$$

Obviously, the first-order system reads now

$$y_1' = y_2, \ \ y_2' = y_3, \ldots, y_{n-1}' = y_n, \ \ y_n' = G(x, y_1, y_2, \ldots, y_n). \qquad (7.4)$$

M. Hermann and M. Saravi, *A First Course in Ordinary Differential Equations: Analytical and Numerical Methods*, DOI 10.1007/978-81-322-1835-7_7, © Springer India 2014

Example 7.1. In Chap. 4, Example 4.23, we have studied the following initial value problem:

$$x''(t) + \alpha^2 x(t) = A \sin(\omega t), \quad x(0) = x'(0) = 0.$$

Formulate this problem as a first-order system.

Solution. We obtain the following system:

$$y_1'(x) = y_2(x),$$
$$y_2'(x) = -\alpha^2 y_1(x) + A \sin(\omega x),$$
$$y_1(0) = y_2(0) = 0.$$

$\square$

In this chapter, we will consider numerical methods for the approximation of solutions of (7.1) which satisfy an additional initial condition $y(x_0) = y_0$, $x_0 \in J$, $y_0 \in \Omega$, i.e., we consider initial value problems (IVPs) of the following form:

$$y'(x) = f(x, y(x)), \quad x \in J,$$
$$y(x_0) = y_0. \tag{7.5}$$

We assume that the function $f(x, y)$ is continuous on the strip

$$S \equiv \{(x, y): x_0 \le x \le x_0 + a, \ y \in \mathbb{R}^n\}$$

and satisfies there a uniform Lipschitz condition

$$\|f(x, y_1) - f(x, y_2)\| \le L \|y_1 - y_2\| \tag{7.6}$$

for all $(x, y_i) \in J \times \Omega$, $i = 1, 2$, and $L \ge 0$. For such a function, we write shortly $f \in \text{Lip}(S)$. Under this assumption, the theorem of Picard and Lindelöf (see, e.g., [17] and Theorem 2.2) guarantees the existence of a unique solution $y(x)$ on the closed interval $[x_0, x_0 + a]$.

7.2 Discretization of an ODE

In most applications it is not possible to represent the exact solution $y(x)$ of (7.5) in closed form. Therefore, $y(x)$ must be approximated by numerical techniques. Let $[x_0, x_N]$ be that interval on which a unique solution of the IVP (7.5) exists. We subdivide $[x_0, x_N]$ into N segments by choosing the points

$$x_0 < x_1 < \cdots < x_N. \tag{7.7}$$

Definition 7.2. The points (7.7) define a *grid*

$$J_h \equiv \{x_0, x_1, \ldots, x_N\}, \tag{7.8}$$

and, therefore, are called *grid points* or *nodes*. Moreover,

$$h_j \equiv x_{j+1} - x_j, \quad j = 0, \ldots, N - 1, \tag{7.9}$$

is the *step-size* from the grid point x_j to the next grid point x_{j+1}. If the step-sizes h_j are constant, i.e., $h_j = h$, $h = (x_N - x_0)/N$, then the grid is called *equidistant*. In the case of a non-equidistant grid, we set

$$h \equiv \max h_j, \quad j = 0, \ldots, N - 1. \tag{7.10}$$

$\square$

Since it is only possible to work with discrete values of the function $y(x)$ on a computer, we need the concept of the discretization of $y(x)$.

Definition 7.3. Under a *discretization* of the function $y(x)$, $x_0 \leq x \leq x_N$, we understand the projection of $y(x)$ onto the underlying grid J_h. The result is a sequence $\{y(x_i)\}_{i=0}^N$, with $x_i \in J_h$. A function which is only defined on a grid is called *grid function*. $\square$

It will be our purpose in the following to compute for the unknown grid function $\{y(x_i)\}_{i=0}^N$ an approximated grid function $\{y_i\}_{i=0}^N$, i.e., approximations y_i for $y(x_i)$, $x_i \in J_h$. This will be realized with a so-called numerical discretization method.

Definition 7.4. A *numerical discretization method* for the approximation of the solution $y(x)$ of the IVP (7.5) is an algorithm which produces a grid function $\{y_i\}_{i=0}^N$, where $y_i \approx y(x_i)$, $x_i \in J_h$. $\square$

Remark 7.5. In order to highlight the dependence of the approximation on the currently used step-size, we will also write y_i^h or $y^h(x_i)$ instead of y_i. $\square$

In this book, we will only consider *one-step methods*.

Definition 7.6. One-step methods are based on one starting value at each integration step (in contrast to linear multistep methods which use the solution values at several previous grid points). Thus, in a typical step of size $h = h_i = x_{i+1} - x_i$, we seek an approximation y_{i+1} to $y(x_{i+1})$, where the result y_i of the previous step is given. We write a (implicit) one-step method in the following form:

$$y_{i+1} = y_i + h_i \, \Phi(x_i, y_i, y_{i+1}; h_i), \quad i = 0, \ldots, N - 1. \tag{7.11}$$

The function Φ is called *increment function* of the method. $\square$

A special class of one-step methods are the *Runge-Kutta methods* which will be discussed in the next section.

7.3   Runge-Kutta Methods

The usual way to develop a discretization method is to represent the ODE (7.2) in
the following equivalent integral form:

$$y(b) = y(a) + \int_a^b f(x, y(x)) \, dx. \tag{7.12}$$

Now, the integrand $f(x, y)$ is approximated by an interpolating polynomial.
Substituting this polynomial into (7.12) makes it very simple to compute the integral
on the right-hand side even if a higher accuracy is demanded.

In the first step, we set $[a, b] \equiv [x_i, x_{i+1}]$. Let us assume that an equidistant
grid with the step-size $h = x_{i+1} - x_i$ is given. In the interval $[x_i, x_{i+1}]$, we
define the following grid points x_{ij} using m mutually distinct numbers $\rho_1, \ldots, \rho_m$,
$0 \le \rho_j \le 1$:

$$x_{ij} \equiv x_i + \rho_j h, \quad j = 1, \ldots, m. \tag{7.13}$$

Let us assume that at these nodes, certain approximations y_{ij}^h for the exact solution
$y(x_{ij})$ of the IVP (7.5) are given and that $\deg(P)$ denotes the degree of a polynomial

$$P(x) = a_0 + a_1 x + \cdots + a_n x^n, \quad a_i \in \mathbb{R}^n.$$

Then one can construct the interpolating polynomial (Lagrangian polynomial)
$P_{m-1}(x)$, with $\deg(P_{m-1}) \le m - 1$, which passes through the points
$(x_{ij}, f(x_{ij}, y_{ij}^h))$. Substituting this polynomial into (7.12) and computing the
integral on the right-hand side, a relation of the following form results:

$$y_{i+1}^h = y_i^h + h \sum_{j=1}^m \beta_j f(x_{ij}, y_{ij}^h). \tag{7.14}$$

Since the vectors y_{ij}^h are not yet known, the above process must be repeated once
again on the m intervals $[a, b] = [x_i, x_{ij}]$, $j = 1, \ldots, m$, using the same grid
$\{x_{ij}\}_{j=1}^m$. We obtain the following equations:

$$y_{ij}^h = y_i^h + h \sum_{l=1}^m \gamma_{il} f(x_{il}, y_{il}^h), \quad j = 1, \ldots, m. \tag{7.15}$$

The formulae (7.14) and (7.15) represent a class of IVP-solvers which belongs to
the more general class of the so-called Runge-Kutta methods.

Today, the class of Runge-Kutta methods is defined more generally as can be
seen in the Definition 7.7. The approach described above should only provide the
motivation for the formulae (7.16).

Fig. 7.1 Butcher diagram of
an m-stage RKM

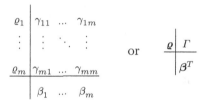

Fig. 7.2 Euler's (forward)
method

$$\begin{array}{c|c} 0 & 0 \\ \hline & 1 \end{array}$$

Definition 7.7. Let $m \in \mathbb{N}$ (*number of stages*). A one-step method of the form

$$y_{i+1}^h = y_i^h + h \sum_{j=1}^m \beta_j k_j,$$

$$k_j = f\left(x_i + \rho_j h, y_i^h + h \sum_{l=1}^m \gamma_{jl} k_l\right), \quad j = 1, \ldots, m, \quad (7.16)$$

is called an m-stage Runge-Kutta method (RKM). The vectors k_j are usually referred to as slopes. Obviously, the free parameters γ_{ij}, ρ_j, and β_j of the RKM must be chosen such that the discretized problem (7.16) converges toward the IVP (7.5) when the step-size h tends to zero. □

Definition 7.8. A Runge-Kutta method (7.16) is called:

- *Explicit* (ERK) iff $\gamma_{ij} = 0$ for $i \leq j$
- *Diagonally implicit* (DIRK) iff $\gamma_{ij} = 0$ for $i < j$
- *Singly diagonally implicit* (SDIRK) iff it is diagonally implicit and in addition it holds $\gamma_{ii} = \gamma$, $\gamma \neq 0$ is a constant
- *Fully implicit* or *implicit* (IRK) iff at least one $\gamma_{ij} \neq 0$ for $i < j$
- *Linearly implicit* (LIRK) iff it is implicit and Newton's method for the solution of the corresponding system of nonlinear algebraic equation is terminated after the first iteration step. □

A Runge-Kutta method can be represented conveniently in a shorthand notation (*Butcher diagram*); see Fig. 7.1.

We now present some of the most important Runge-Kutta methods:

1. Euler's (forward) method: $\quad y_{i+1}^h = y_i^h + h f(x_i, y_i^h)$
   This method has been proposed by the famous mathematician Leonard Euler (see [12]) and is one of the most frequently discussed methods. Since we can write $y_{i+1}^h = y_i^h + h \cdot 1 \cdot k_1$, with $k_1 = f(x_i + 0 \cdot h, y_i^h + h \cdot 0 \cdot k_1)$, the Butcher diagram presented in Fig. 7.2 results.
2. Euler's (backward) method: $\quad y_{i+1}^h = y_i^h + h f(x_{i+1}, y_{i+1}^h)$

Fig. 7.3 Euler's (backward) method

1	1
	1

Fig. 7.4 Trapezoidal method

0	0	0
1	$\frac{1}{2}$	$\frac{1}{2}$
	$\frac{1}{2}$	$\frac{1}{2}$

Fig. 7.5 Heun's method

0		
1	1	0
	$\frac{1}{2}$	$\frac{1}{2}$

We can write $y_{i+1}^h = y_i^h + h \cdot 1 \cdot k_1$, with $k_1 = f(x_i + 1 \cdot h, y_i^h + h \cdot 1 \cdot k_1)$. This yields the Butcher diagram which is shown in Fig. 7.3.

3. Trapezoidal method:　$y_{i+1}^h = y_i^h + \frac{h}{2} f(x_i, y_i^h) + \frac{h}{2} f(x_{i+1}, y_{i+1}^h)$

We have $y_{i+1}^h = y_i^h + h\{\frac{1}{2}k_1 + \frac{1}{2}k_2\}$, with

$k_1 = f(x_i + 0 \cdot h, y_i^h + h(0 \cdot k_1 + 0 \cdot k_2))$ and

$k_2 = f(x_i + 1 \cdot h, y_i^h + h(\frac{1}{2} \cdot k_1 + \frac{1}{2} \cdot k_2))$.

The corresponding Butcher diagram is given in Fig. 7.4.

4. Heun's method:　$y_{i+1}^h = y_i^h + h\{\frac{1}{2} f(x_i, y_i^h) + \frac{1}{2} f(x_{i+1}, y_i^h + hf(x_i, y_i^h))\}$

Replacing y_{i+1}^h by Euler's (forward) formula on the right-hand side of the trapezoidal method gives the so-called Heun's method which has been proposed by [21].

Here, we have $y_{i+1}^h = y_i^h + h\{\frac{1}{2}k_1 + \frac{1}{2}k_2\}$,　with

$$k_1 = f(x_i + 0 \cdot h, y_i^h + h\{0 \cdot k_1 + 0 \cdot k_2\}),$$

$$k_2 = f(x_i + 1 \cdot h, y_i^h + h\{1 \cdot k_1 + 0 \cdot k_2\}).$$

The corresponding Butcher diagram is given in Fig. 7.5. From now on, in the Butcher diagrams we omit the upper triangle of the matrix Γ (it consists only of zeros) when the method is explicit.

Fig. 7.6 Classical
Runge-Kutta method

$$
\begin{array}{c|cccc}
0 & & & & \\
\dfrac{1}{2} & \dfrac{1}{2} & & & \\
\dfrac{1}{2} & 0 & \dfrac{1}{2} & & \\
1 & 0 & 0 & 1 & \\
\hline
 & \dfrac{1}{6} & \dfrac{1}{3} & \dfrac{1}{3} & \dfrac{1}{6}
\end{array}
$$

5. A widely used method is the classical Runge-Kutta method. It has been developed by Runge [24] and was formulated in the currently conventional form by Kutta [23]. The corresponding Butcher diagram is given in Fig. 7.6.

   Substituting the parameters given in Fig. 7.6 into the formula (7.16) of a general Runge-Kutta method, we obtain the well-known procedure:

$$
y_{i+1}^h = y_i^h + h\{\frac{1}{6}k_1 + \frac{1}{3}k_2 + \frac{1}{3}k_3 + \frac{1}{6}k_4\},
$$

$$
k_1 = f(x_i, y_i^h), \quad k_2 = f(x_i + \frac{h}{2}, y_i^h + \frac{h}{2}k_1),
$$

$$
k_3 = f(x_i + \frac{h}{2}, y_i^h + \frac{h}{2}k_2), \quad k_4 = f(x_i + h, y_i^h + hk_3). \tag{7.17}
$$

6. Embedded Runge-Kutta methods are composed of a pair of formulae of orders p and q (usually $q = p + 1$) that share the same function evaluation points. A frequently used class is the Runge-Kutta-Fehlberg family (see, e.g., [13, 14, 16]). The parameters of the Runge-Kutta-Fehlberg 4(5) pair are given in Fig. 7.7. It can be seen immediately that the matrix Γ of RKF4 is a leading main submatrix of those of RKF5.

7. An important class of embedded methods is the Runge-Kutta-Verner family (see [27]). The parameters of the Runge-Kutta-Verner 5(6) pair are given in Fig. 7.8.

8. Another very frequently used class of embedded methods is the Runge-Kutta-Dormand-Prince family (see [10]). This family is designed to minimize the local error in the approximation of the higher-order method of the respective pair, in anticipation that the latter will be used for the next step. The 4(5) pair given in Fig. 7.9 has seven stages, but the last stage is the same as the first stage for the next step, so this method has the cost of a six-stage method.

Example 7.9. In Example 3.36, we have studied the undamped oscillatory motion of a mass-spring combination governed by the initial value problem:

$$
y''(x) + \lambda y(x) = 0, \quad y(0) = y_0, \quad y'(0) = v_0.
$$

0						
$\dfrac{1}{4}$	$\dfrac{1}{4}$					
$\dfrac{3}{8}$	$\dfrac{3}{32}$	$\dfrac{9}{32}$				
$\dfrac{12}{13}$	$\dfrac{1932}{2197}$	$-\dfrac{7200}{2197}$	$\dfrac{7296}{2197}$			
1	$\dfrac{439}{216}$	-8	$\dfrac{3680}{513}$	$-\dfrac{845}{4104}$		
$\dfrac{1}{2}$	$-\dfrac{8}{27}$	2	$-\dfrac{3544}{2565}$	$\dfrac{1859}{4104}$	$-\dfrac{11}{40}$	
RKF4	$\dfrac{25}{216}$	0	$\dfrac{1408}{2565}$	$\dfrac{2197}{4104}$	$-\dfrac{1}{5}$	
RKF5	$\dfrac{16}{135}$	0	$\dfrac{6656}{12825}$	$\dfrac{28561}{56430}$	$-\dfrac{9}{50}$	$\dfrac{2}{55}$

Fig. 7.7 Runge-Kutta-Fehlberg 4(5)

The corresponding system of two first-order equations is

$$y_1' = y_2,$$
$$y_2' = -\lambda y_1,$$
$$y_1(0) = y_0, \quad y_2(0) = v_0,$$

or in matrix-vector notation

$$\begin{pmatrix} y_1' \\ y_2' \end{pmatrix} = \begin{pmatrix} 0 & 1 \\ -\lambda & 0 \end{pmatrix} \begin{pmatrix} y_1 \\ y_2 \end{pmatrix}, \quad \begin{pmatrix} y_1(0) \\ y_2(0) \end{pmatrix} = \begin{pmatrix} y_0 \\ v_0 \end{pmatrix}.$$

The exact solution (see Example 3.36) is

$$y_1(x) = y_0 \cos(\sqrt{\lambda}x) + \frac{v_0}{\sqrt{\lambda}} \sin(\sqrt{\lambda}x),$$

$$y_2(x) = -y_0 \sqrt{\lambda} \sin(\sqrt{\lambda}x) + v_0 \cos(\sqrt{\lambda}x).$$

For this problem, Euler's (forward) method $y_{i+1}^h = y_i^h + h f(x_i, y_i^h)$ reads

$$\begin{pmatrix} y_{1,i+1}^h \\ y_{2,i+1}^h \end{pmatrix} = \begin{pmatrix} y_{1,i}^h \\ y_{2,i}^h \end{pmatrix} + h \begin{pmatrix} 0 & 1 \\ -\lambda & 0 \end{pmatrix} \begin{pmatrix} y_{1,i}^h \\ y_{2,i}^h \end{pmatrix}, \quad i = 0, 1, \ldots.$$

0								
$\dfrac{1}{18}$	$\dfrac{1}{18}$							
$\dfrac{1}{6}$	$-\dfrac{1}{12}$	$\dfrac{1}{4}$						
$\dfrac{2}{9}$	$-\dfrac{2}{81}$	$\dfrac{4}{27}$	$\dfrac{8}{81}$					
$\dfrac{2}{3}$	$\dfrac{40}{33}$	$-\dfrac{4}{11}$	$-\dfrac{56}{11}$	$\dfrac{54}{11}$				
1	$-\dfrac{369}{73}$	$\dfrac{72}{73}$	$\dfrac{5380}{219}$	$-\dfrac{12285}{584}$	$\dfrac{2695}{1752}$			
$\dfrac{8}{9}$	$-\dfrac{8716}{891}$	$\dfrac{656}{297}$	$\dfrac{39520}{891}$	$-\dfrac{416}{11}$	$\dfrac{52}{27}$	0		
1	$\dfrac{3015}{256}$	$-\dfrac{9}{4}$	$-\dfrac{4219}{78}$	$\dfrac{5985}{128}$	$-\dfrac{539}{384}$	0	$\dfrac{693}{3328}$	
RKV5	$\dfrac{3}{80}$	0	$\dfrac{4}{25}$	$\dfrac{243}{1120}$	$\dfrac{77}{160}$	$\dfrac{73}{700}$	0	
RKV6	$\dfrac{57}{640}$	0	$-\dfrac{16}{65}$	$\dfrac{1377}{2240}$	$\dfrac{121}{320}$	0	$\dfrac{891}{8320}$	$\dfrac{2}{35}$

Fig. 7.8 Runge-Kutta-Verner 5(6)

Let us choose $\lambda = 50$, $y_0 = 1$, $v_0 = 2$ and the step-size $h = 0.01$. For the exact solution, we get

$$y_1(0.01) = 1.0174844, \quad y_2(0.01) = 1.4954186.$$

Executing one step of Euler's (forward) method, we evaluate the following approximations:

$$y_{1,1}^{0.01} = 1.0200000, \quad y_{2,1}^{0.01} = 1.5000000.$$

$\square$

Example 7.10. Let us solve the initial value problem studied in Example 7.9 with Heun's method

$$y_{i+1}^h = y_i^h + h\{\tfrac{1}{2}f(x_i, y_i^h) + \tfrac{1}{2}f(x_{i+1}, y_i^h + hf(x_i, y_i^h))\},$$

0						
$\dfrac{1}{5}$	$\dfrac{1}{5}$					
$\dfrac{3}{10}$	$\dfrac{3}{40}$	$\dfrac{9}{40}$				
$\dfrac{4}{5}$	$\dfrac{44}{45}$	$-\dfrac{56}{15}$	$\dfrac{32}{9}$			
$\dfrac{8}{9}$	$\dfrac{19372}{6561}$	$-\dfrac{25360}{2187}$	$\dfrac{64448}{6561}$	$-\dfrac{212}{729}$		
1	$\dfrac{9017}{3168}$	$-\dfrac{355}{33}$	$\dfrac{46732}{5247}$	$\dfrac{49}{176}$	$-\dfrac{5103}{18656}$	
1	$\dfrac{35}{384}$	0	$\dfrac{500}{1113}$	$\dfrac{125}{192}$	$-\dfrac{2187}{6748}$	$\dfrac{11}{84}$
	$\dfrac{35}{384}$	0	$\dfrac{500}{1113}$	$\dfrac{125}{192}$	$-\dfrac{2187}{6784}$	$\dfrac{11}{84}$	0
	$\dfrac{5179}{57600}$	0	$\dfrac{7571}{16695}$	$\dfrac{393}{640}$	$-\dfrac{92097}{339200}$	$\dfrac{187}{2100}$	$\dfrac{1}{40}$

Fig. 7.9 Runge-Kutta-Dormand-Prince 4(5)

i.e.,

$$
\begin{pmatrix} y^{h}_{1,i+1} \\ y^{h}_{2,i+1} \end{pmatrix} = \begin{pmatrix} y^{h}_{1,i} \\ y^{h}_{2,i} \end{pmatrix} + h \left\{ \frac{1}{2} \begin{pmatrix} 0 & 1 \\ -\lambda & 0 \end{pmatrix} \begin{pmatrix} y^{h}_{1,i} \\ y^{h}_{2,i} \end{pmatrix} \right.
$$
$$
\left. + \frac{1}{2} \begin{pmatrix} 0 & 1 \\ -\lambda & 0 \end{pmatrix} \left[\begin{pmatrix} y^{h}_{1,i} \\ y^{h}_{2,i} \end{pmatrix} + h \begin{pmatrix} 0 & 1 \\ -\lambda & 0 \end{pmatrix} \begin{pmatrix} y^{h}_{1,i} \\ y^{h}_{2,i} \end{pmatrix} \right] \right\}.
$$

With the step-size $h = 0.01$, we obtain in the first integration step the approximations

$$
y^{0.01}_{1,1} = 1.017500, \quad y^{0.01}_{2,1} = 1.495000.
$$

This shows that Heun's method produces better results than Euler's (forward) method. In the next sections, we will show why this is the case. □

Example 7.11. Let us consider the same initial value problem as in the previous two examples. We integrate this problem with three different numerical methods on the interval $[0, 5]$. The step-size is $h = 1/12$ and the IVP-solvers are Euler's (forward) method, Heun's method, and the Runge-Kutta-Fehlberg method (RKF3) which is defined by the Butcher diagram given in Fig. 7.10. The numerical results

Fig. 7.10 3-step
Runge-Kutta-Fehlberg
method

$$
\begin{array}{c|ccc}
0 & & & \\
1 & 1 & & \\
\dfrac{1}{2} & \dfrac{1}{4} & \dfrac{1}{4} & \\
\hline
 & \dfrac{1}{6} & \dfrac{1}{6} & \dfrac{4}{6}
\end{array}
$$

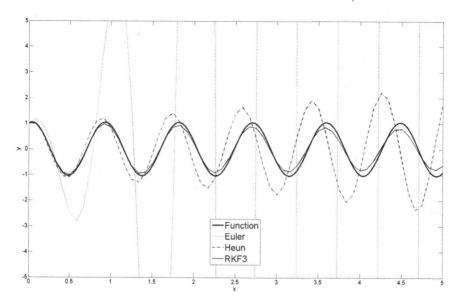

Fig. 7.11 Comparison of three different IVP-solvers

are visualized in Fig. 7.11. It can be seen that the 3-stage Runge-Kutta-Fehlberg
method produces the best results, whereas Euler's (forward) method goes out of the
range of the picture, i.e., the error becomes large very quickly.   □

7.4   Local Truncation Error and Consistency

Given an equidistant grid J_h, let us consider the (implicit) one-step method

$$
y_{i+1}^h = y_i^h + h\,\boldsymbol{\Phi}(x_i, y_i^h, y_{i+1}^h; h). \tag{7.18}
$$

For any grid function u on J_h, we define the difference operator $\triangle_h$ as

$$
\triangle_h\, u(x_{i+1}) \equiv \frac{1}{h}\left[u(x_{i+1}) - u(x_i) - h\,\boldsymbol{\Phi}(x_i, u(x_i), u(x_{i+1}); h)\right]. \tag{7.19}
$$

If y^h denotes the grid function which is defined by (7.18), we have

$$\triangle_h \, y^h(x_{i+1}) = 0.$$

Much of the study of numerical ODEs is concerned with the errors that occur at each step when the difference equation (7.18) is used instead of the given IVP (7.5), and how they accumulate. A measure of the error that arises at each step is the *local truncation error*.

Definition 7.12. The local truncation error $\delta(\cdot)$ is the residual of the difference operator $\triangle_h$ when it is applied to the exact solution of the IVP (7.5):

$$\delta(x_{i+1}, y(t_{i+1}); h) \equiv \triangle_h \, y(x_{i+1})$$

$$= \frac{1}{h} \left[y(x_{i+1}) - y(x_i) - h \, \boldsymbol{\Phi}(x_i, y(x_i), y(x_{i+1}); h) \right]. \qquad (7.20)$$

□

The local truncation error measures how closely the difference operator approximates the differential operator. It turns out that the basic requirement for a one-step method is its consistency. This property is defined by the local truncation error.

Definition 7.13. The one-step method (7.18) is said to be consistent of order p (with the IVP) if p is the largest positive integer such that the local truncation error satisfies

$$\delta(x_{i+1}, y(x_{i+1}); h) = O(h^p), \qquad h \to 0, \qquad (7.21)$$

where $O(h^p)$ implies the existence of finite constants C and $h_0 > 0$ such that

$$\delta(x_{i+1}, y(x_{i+1}); h) \leq C \, h^p$$

for all $h \leq h_0$. Consistency normally implies that the order $p \geq 1$.   □

Remark 7.14. The consistency guarantees that the numerical one-step method (discretized problem) converges to the given IVP (continuous problem) if the step-size h goes to 0.   □

Remark 7.15. The order of consistency is the answer to the question which we have formulated in Examples 7.9 and 7.10, namely, the larger the order of consistency p, the quicker the one-step method converges to the IVP. More precisely, for a one-step method whose order is larger, the step-size h can be chosen larger than for a method with a smaller order p.   □

If the exact solution is sufficiently smooth, we can find the order of consistency of a one-step method by expanding the exact solution $y(x_{i+1}) = y(x_i + h)$ and $y(x_i) = y(x_{i+1} - h)$, respectively, and the corresponding incremental function $\boldsymbol{\Phi}$

in Taylor series and substituting these expansions into (7.20). By identifying the terms having the same power of h in the resulting formula, we can say that the order equals p if the coefficients of h^0 up to h^p vanish.

For simplicity, we shall assume f to be autonomous, i.e., $f(x, y) = f(y)$, which avoids partial derivatives w.r.t. x. Otherwise, (7.1) can be transformed by a so-called autonomization into an autonomous system of dimension $n+1$ as follows: Setting $w = (w_1, w_2, \ldots, w_{n+1})^T \equiv (x, y_1, y_2, \ldots, y_n)^T$ and adding the trivial differential equation $\dfrac{d}{dx} x = 1$ to the system (7.1), we get

$$
\begin{aligned}
w_1' &= 1 \\
w_2' &= f_1(w_1, \ldots, w_{n+1}) \\
&\vdots = \vdots \\
w_{n+1}' &= f_n(w_1, \ldots, w_{n+1})
\end{aligned}
\qquad \Longleftrightarrow \qquad w' = g(w). \qquad (7.22)
$$

Obviously, the ODE (7.22) is autonomous, and we can postulate this assumption without loss of generality. To simplify the representation, we suppose that the problem is scalar, i.e., $f(y) \in \mathbb{R}$. The properties of the methods considered in Example 7.1 keep their validity also with n-dimensional IVPs of the form (7.5).

Under our assumptions, we have

$$
y(x_{i+1}) = y(x_i + h)
$$

$$
= y(x_i) + h y'(x_i) + \frac{h^2}{2} y''(x_i) + \frac{h^3}{6} y'''(x_i) + O(h^4),
$$

$$
y(x_i) = y(x_{i+1} - h)
$$

$$
= y(x_{i+1}) - h y'(x_i) + \frac{h^2}{2} y''(x_i) - \frac{h^3}{6} y'''(x_i) + O(h^4).
$$

Since $y'(t)$ is correlated with $f(y)$ by the ODE, i.e., $y' = f(y)$, the higher-order derivatives of $y(x)$ can be expressed in terms of $f(y)$ and its derivatives. One obtains

$$
\begin{aligned}
y' &= f(y) \equiv f \\
y'' &= f_y y' = f_y f \quad \left(f_y \equiv \frac{d f(y)}{d y}, \text{ etc.} \right) \\
y'' &= f_{yy} f^2 + (f_y)^2 f \\
&\vdots
\end{aligned}
$$

Now we want to determine for some of the already described one-step methods the corresponding order of consistency. For this, we use the following abbreviations:

$$
f_i \equiv f(y(x_i)), \quad (f_i)_y \equiv \frac{d f}{d y}(y(x_i)), \quad (f_i)_{yy} \equiv \frac{d^2 f}{d y^2}(y(x_i)), \quad \text{etc.}
$$

1. Euler's (forward) method: $y_{i+1}^h = y_i^h + hf(y_{i+1}^h)$

$$\delta(\cdot) = \frac{1}{h}\left[y(x_{i+1}) - y(x_i) - hf(y(x_{i+1}))\right]$$

$$= \frac{1}{h}\left[hy'(x_{i+1}) + O(h^2) - hf(y(x_{i+1}))\right]$$

$$= \frac{1}{h}\left[hf(y(x_{i+1})) + O(h^2) - hf(y(x_{i+1}))\right] = O(h^1)$$

Thus, the method has the order of consistency $p = 1$.

2. Trapezoidal method: $y_{i+1}^h = y_i^h + h\left[\frac{1}{2}f(y_i^h) + \frac{1}{2}f(y_{i+1}^h)\right]$

$$\delta(\cdot) = \frac{1}{h}\left[y(x_{i+1}) - y(x_i) - h\left(\frac{1}{2}f(y(x_i)) + \frac{1}{2}f(y(x_{i+1}))\right)\right]$$

$$= \frac{1}{h}\left[hy'(x_i) + \frac{h^2}{2}y''(x_i) + \frac{h^3}{6}y'''(x_i) + O(h^4)\right.$$

$$\left.-h\left(\frac{1}{2}f(y(x_i)) + \frac{1}{2}f(y(x_{i+1}))\right)\right]$$

We obtain

$$\delta(\cdot) = \frac{1}{h}\left[hf_i + \frac{h^2}{2}(f_i)_y f_i + \frac{h^3}{6}\left((f_i)_{yy} f_i^2 + (f_i)_y^2 f_i\right) + O(h^4)\right.$$

$$\left.-\frac{h}{2}(f_i + f_{i+1})\right]$$

It is

$$f_{i+1} = f(y(x_{i+1})) = f(y(x_i + h))$$

$$= f\left(y(x_i) + \underbrace{hf_i + \frac{h^2}{2}(f_i)_y f_i + O(h^3)}_{\equiv \tilde{h}}\right)$$

$$= f(y(x_i) + \tilde{h}) = f_i + (f_i)_y \tilde{h} + \frac{1}{2}(f_i)_{yy}\tilde{h}^2 + \cdots$$

$$= f_i + h(f_i)_y f_i + \frac{h^2}{2}(f_i)_y^2 f_i + \frac{h^2}{2}(f_i)_{yy} f_i^2 + O(h^3).$$

Fig. 7.12 A 3-stage RKM

$$
\begin{array}{c|ccc}
0 & & & \\
\dfrac{1}{2} & \dfrac{1}{2} & & \\
1 & -1 & 2 & \\
\hline
 & \dfrac{1}{6} & \dfrac{2}{3} & \dfrac{1}{6}
\end{array}
$$

Substituting this expression into $\delta(\cdot)$, we obtain

$$
\delta(\cdot) = \frac{1}{h}\left[hf_i + \frac{h^2}{2}(f_i)_y f_i + \frac{h^3}{6}\left((f_i)_{yy} f_i^2 + (f_i)_y^2 f_i \right) \right.
$$
$$
\left. - hf_i - \frac{h^2}{2}(f_i)_y f_i - \frac{h^3}{4}\left((f_i)_{yy} f_i^2 + (f_i)_y^2 f_i \right) + O(h^4) \right]
$$
$$
= O(h^2).
$$

Thus, the method has the order of consistency $p = 2$.

3. Let us now consider the 3-stage Runge-Kutta method which is defined by the Butcher diagram given in Fig. 7.12.

Looking at (7.16), we have to substitute the coefficients presented in the diagram into the formulae

$$
y_{i+1}^h = y_i^h + h \sum_{j=1}^{3} \beta_j k_j, \quad k_j \equiv f\left(y_i^h + h \sum_{l=1}^{3} \gamma_{jl} k_l \right), \quad j = 1, 2, 3.
$$

To determine the local truncation error of this method, the exact solution is substituted into k_1, k_2, and k_3. Then, the resulting functions are expanded in Taylor series:

$$
k_1 = f_i,
$$
$$
k_2 = f\left(y(x_i) + \frac{h}{2} f_i \right)
$$
$$
= f_i + \frac{h}{2}(f_i)_y f_i + \frac{h^2}{8}(f_i)_{yy} f_i^2 + \frac{h^3}{48}(f_i)_{yyy} f_i^3 + O(h^4),
$$
$$
k_3 = f\left(y(x_i) + h[-k_1 + 2k_2] \right)
$$
$$
= f\left(y(x_i) + hf_i + h^2(f_i)_y f_i + \frac{h^3}{4}(f_i)_{yy} f_i^2 + O(h^4) \right)
$$
$$
= f_i + (f_i)_x \left\{ hf_i + h^2(f_i)_y f_i + \frac{h^3}{4}(f_i)_{yy} f_i^2 + O(h^4) \right\}
$$

$$+ \frac{1}{2}(f_i)_{yy} \left\{ h^2 f_i^2 + 2h^3 (f_i)_y f_i^2 + O(h^4) \right\}$$

$$+ \frac{1}{6}(f_i)_{yyy} \left\{ h^3 f_i^3 + O(h^4) \right\} + O(h^4)$$

$$= f_i + h(f_i)_y f_i + h^2 \left\{ (f_i)_y^2 f_i + \frac{1}{2}(f_i)_{yy} f_i^2 \right\}$$

$$+ h^3 \left\{ \frac{5}{4}(f_i)_{yy}(f_i)_y f_i^2 + \frac{1}{6}(f_i)_{yyy} f_i^3 \right\} + O(h^4).$$

Thus, Φ can be written as

$$h \Phi(\cdot) = h \sum_{j=1}^{3} \beta_j k_j = h\frac{1}{6} f_i + h\frac{2}{3} f_i + \frac{h^2}{3}(f_i)_y f_i + \frac{h^3}{12}(f_i)_{yy} f_i^2$$

$$+ \frac{h}{6} f_i + \frac{h^2}{6}(f_i)_y f_i + \frac{h^3}{6}(f_i)_y^2 f_i + \frac{h^3}{12}(f_i)_{yy} f_i^2 + O(h^4)$$

$$= hf_i + \frac{h^2}{2}(f_i)_y f_i + \frac{h^3}{6} \left\{ (f_i)_y^2 f_i + (f_i)_{yy} f_i^2 \right\} + O(h^4).$$

Substituting these expansions into $\delta(\cdot)$, we get

$$\delta(\cdot) = \frac{1}{h} \left[hf_i + \frac{h^2}{2}(f_i)_y f_i + \frac{h^3}{6} \left\{ (f_i)_{yy} f_i^2 + (f_i)_y^2 f_i \right\} + O(h^4) \right.$$

$$\left. - hf_i - \frac{h^2}{2}(f_i)_y f_i - \frac{h^3}{6} \left\{ (f_i)_{yy} f_i^2 + (f_i)_y^2 f_i \right\} + O(h^4) \right]$$

$$= O(h^3).$$

We can now deduce that the order of consistency is at least $p = 3$. To show that the method has exactly the order of consistency 3, we must use more terms of the Taylor expansions.

4. In analogous manner, it can be shown that the 5- and 6-stage Runge-Kutta-Fehlberg methods RKF4 and RKF5 (see Fig. 7.7) have the order of consistency 4 and 5, respectively. This shows that the number of stages m and the order of consistency p must not agree. In Remark 7.16, the reader finds a statement about this problem.

Remark 7.16. Let m be the number of stages of an RKM for the IVP (7.5). In the papers [4, 5] Butcher has shown that for $p \geq 5$, no explicit Runge-Kutta method exists of order p with $m = p$.

Adding an additional stage increases the computational costs. Thus, the explicit Runge-Kutta formulae with $m = 4$ are optimal. An example is the classical Runge-Kutta method (7.17). Looking at Table 7.1, we conclude that RKF4 is not optimal, whereas RKF5 is again optimal.

Table 7.1 Minimum number of stages of explicit RKM

p	1	2	3	4	5	6	7	8	$\cdots$	≥ 9
m_p	1	2	3	4	6	7	9	11	$\cdots$	$\geq p+3$

The values m_p are called *Butcher barriers*. The relation of consistency order p to the number of required stages m gets worse with growing order. $\quad\square$

Let us return to the Runge-Kutta methods (7.16) which obviously belong to the class of one-step methods. Here, we have

$$\boldsymbol{\Phi}(x_i, y_i^h, y_{i+1}^h; h) = \sum_{j=1}^{m} \beta_j \boldsymbol{k}_j. \tag{7.23}$$

In Definition 7.7, nothing is stated about the question how the coefficients γ_{ij}, ϱ_j, and β_j have to be chosen. As we have seen above, the minimum requirement on a one-step method is its consistency. The following relations exist between the coefficients in the Butcher diagram and the consistency of a Runge-Kutta method.

Theorem 7.17. *The m-stage Runge-Kutta method (7.16) is consistent with the IVP (7.5) iff*

$$\beta_1 + \beta_2 + \cdots + \beta_m = 1. \tag{7.24}$$

Proof. The statement follows immediately from a Taylor expansion of the $\boldsymbol{k}_j$'s:

$$\boldsymbol{k}_j = \boldsymbol{f}\left(x_i + \varrho_j h, \boldsymbol{y}(x_i) + h \sum_{l=1}^{m} \gamma_{jl} \boldsymbol{k}_l\right) = \boldsymbol{f}(x_i, \boldsymbol{y}(x_i)) + O(h).$$

Substituting this expansion into the incremental function $\boldsymbol{\Phi}$ we get

$$\boldsymbol{\Phi}(\cdot) = \sum_{l=1}^{m} \beta_l \boldsymbol{k}_l = \boldsymbol{f}(x_i, \boldsymbol{y}(x_i)) \sum_{l=1}^{m} \beta_l + O(h).$$

Now, the local truncation error reads

$$\boldsymbol{\delta}(\cdot) = \frac{1}{h}\left[h \boldsymbol{f} - h \boldsymbol{f} \sum_{l=1}^{m} \beta_l + O(h^2)\right].$$

Consequently $\boldsymbol{\delta}(\cdot) = O(h)$ is valid iff Eq. (7.24) is satisfied. $\quad\blacksquare$

By means of formula (7.22), we have seen how from an n-dimensional system of nonautonomous ODEs an equivalent $(n+1)$-dimensional system of autonomous

ODEs can be generated. The question now arises whether a special method produces the same results if it is applied to the original and to the autonomized problem. An answer to that question gives the next theorem.

Theorem 7.18. *An ERK is invariant up to an autonomization iff it is consistent and the following relation holds:*

$$\varrho_j = \sum_{l=1}^{m} \gamma_{jl}, \quad j = 1, \ldots, m. \tag{7.25}$$

Proof. See, e.g., the book [19].   ∎

7.5   Construction of Runge-Kutta Methods

In this section we will give an answer to the question of how a Runge-Kutta method can be constructed with a prescribed order of consistency p. This construction is based on two steps:

- Developing equations which constitute conditions on the coefficients ϱ, Γ, and β such that the resulting Runge-Kutta method has the order of consistency p.
- Solving this (under-determined) system of nonlinear algebraic equations, i.e., the indication of relevant sets of coefficients. The coefficients should be given exactly as rational expressions.

We will demonstrate the first step for the order of consistency $p = 4$. Consider the system of n first-order autonomous ODEs:

$$\dot{y} = f(y). \tag{7.26}$$

To compute the Taylor expansion of the exact solution $y(x)$, we require the following relations between the derivatives of $y(x)$ and the function $f(y)$:

$$\dot{y} = f(y) \equiv f$$
$$\ddot{y} = f_y f$$
$$\dddot{y} = f_{yy}(f, f) + f_y f_y f$$
$$\ddddot{y} = f_{yyy}(f, f, f) + 3f_{yy}(f_y f, f) + f_y f_{yy}(f, f) + f_y f_y f_y f$$
$$\vdots \tag{7.27}$$

One recognizes that the continuation of this process, although theoretically clear, soon leads to very complicated expressions. It is therefore advantageous to use a graphical representation which is known as the theory of rooted trees (see, e.g., [3, 5, 15]). However, for our purposes, it is sufficient to have the formulae (7.27) at hand.

Using the abbreviation $f^i \equiv f(y(x_i))$, we obtain the following Taylor expansion for $y(x_{i+1})$:

$$y(x_{i+1}) = y(x_i) + hf^i + \frac{h^2}{2!}f_y^i f^i + \frac{h^3}{3!}\left[f_{yy}^i(f^i, f^i) + f_y^i f_y^i f^i\right]$$

$$+ \frac{h^4}{4!}\left[f_{yyy}^i(f^i, f^i, f^i) + 3f_{yy}^i(f_y^i f^i, f^i)\right.$$

$$\left. + f_y^i f_{yy}^i(f^i, f^i) + f_y^i f_y^i f_y^i f^i\right] + O(h^5). \quad (7.28)$$

In order to represent the incremental function Φ in powers of h, too, we do not use the Taylor expansion technique, since the continued differentiation quickly leads to confusing expressions. We use a totally different strategy which is known as *boot-strapping process*. One can, namely, utilize the fact that in the recursive equations,

$$k_j = f\left(y(x_i) + h\sum_{l=1}^m \gamma_{jl} k_l\right), \quad j = 1, \ldots, m, \quad (7.29)$$

the stages k_j within the function f are multiplied by h. From the continuity of f, we deduce

$$k_j = O(1) \quad \text{for} \quad j = 1, \ldots, m, \quad h \to 0.$$

Using this information in the right-hand side of (7.29), we obtain

$$k_j = f(y(x_i) + O(h)) = f^i + O(h), \quad j = 1, \ldots, m. \quad (7.30)$$

In the next step, this new information (7.30) is again substituted into (7.29). It follows

$$k_j = f\left(y(x_i) + h\sum_l \gamma_{jl} f^i + O(h^2)\right) = f^i + h\varrho_j f_y^i f^i + O(h^2), \quad (7.31)$$

where $\varrho_j \equiv \sum_j \gamma_{jl}$. The third step yields

$$k_j = f\left(y(x_i) + h\sum_l \gamma_{jl}\left(f^i + h\varrho_l f_y^i f^i\right) + O(h^3)\right)$$

$$= f^i + h\varrho_j f_y^i f^i + h^2 \sum_l \gamma_{jl}\varrho_l f_y^i f_y^i f^i + \frac{h^2}{2}\varrho_j^2 f_{yy}^i(f^i, f^i) + O(h^3).$$

$$(7.32)$$

Finally, in the last (fourth) step one gets

$$
\begin{aligned}
k_j = f &\left(y(x_i) + h\varrho_j f^i + h^2 \sum_l \gamma_{jl}\varrho_l f^i_y f^i + h^3 \sum_l \sum_r \gamma_{jl}\gamma_{lr}\varrho_r f^i_y f^i_y f^i \right.\\
&\left. + \frac{h^3}{2} \sum_l \gamma_{jl}\varrho_l^2 f^i_{yy}(f^i, f^i) + O(h^4) \right)\\
= f^i &+ h\varrho_j f^i_y f^i + h^2 \sum_l \gamma_{jl}\varrho_l f^i_y f^i_y f^i + \frac{h^2}{2}\varrho_j^2 f^i_{yy}(f^i, f^i)\\
&+ h^3 \sum_l \sum_r \gamma_{jl}\gamma_{lr}\varrho_r f^i_y f^i_y f^i_y f^i + \frac{h^3}{2} \sum_l \gamma_{jl}\varrho_l^2 f^i_y f^i_{yy}(f^i, f^i)\\
&+ h^3 \sum_l \varrho_l \gamma_{jl}\varrho_j f^i_{yy}(f^i_y f^i, f^i) + \frac{h^3}{6}\varrho_j^3 f^i_{yyy}(f^i, f^i, f^i) + O(h^4).
\end{aligned}
\tag{7.33}
$$

In this recursive manner, we obtain step-by-step new information. Using (7.33) in the expression $h\,\boldsymbol{\Phi}(\cdot)$, we get

$$
\begin{aligned}
h\,\boldsymbol{\Phi}(\cdot) = h &\sum_j \beta_j f^i + h^2 \sum_j \beta_j \varrho_j f^i_y f^i\\
&+ \frac{h^3}{3!}\left(3\sum_j \beta_j \varrho_j^2 f^i_{yy}(f^i, f^i) + 6\sum_j \sum_l \beta_j \gamma_{jl}\varrho_l f^i_y f^i_y f^i \right)\\
&+ \frac{h^4}{4!}\left(4\sum_j \beta_j \varrho_j^3 f^i_{yyy}(f^i, f^i, f^i) \right.\\
&\qquad\quad + 24 \sum_j \sum_l \beta_j \varrho_j \gamma_{jl}\varrho_l f^i_{yy}(f^i_y f^i, f^i)\\
&\qquad\quad + 12 \sum_j \sum_l \beta_j \gamma_{jl}\varrho_l^2 f^i_y f^i_{yy}(f^i, f^i)\\
&\qquad\quad \left. + 24 \sum_j \sum_l \sum_r \beta_j \gamma_{jl}\gamma_{lr}\varrho_r f^i_y f^i_y f^i_y f^i \right) + O(h^5).
\end{aligned}
\tag{7.34}
$$

Substituting (7.28) and (7.34) into the formula (7.20) of the local truncation error $\delta(\cdot)$ and claiming that the expressions containing h, h^2, h^3, and h^4, respectively, are equal to zero, we obtain the statement of Theorem 7.19. In particular, we get conditions on the free coefficients of the Runge-Kutta method which guarantee a certain order of consistency. These conditions are called *order conditions*.

Table 7.2 Number of order conditions N_p for the Runge-Kutta methods

p	1	2	3	4	5	6	7	8	9	10	20
N_p	1	2	4	8	17	37	85	200	486	1,205	20,247,374

Theorem 7.19. *Let f be a sufficiently smooth function. A Runge-Kutta method has the order of consistency $p = 1$ iff the coefficients satisfy*

$$\sum_j \beta_j = 1; \tag{7.35}$$

the order of consistency $p = 2$ iff in addition the condition

$$\sum_j \beta_j \varrho_j = \frac{1}{2} \tag{7.36}$$

is satisfied; the order of consistency $p = 3$ iff in addition the two conditions

$$\sum_j \beta_j \varrho_j^2 = \frac{1}{3}, \quad \sum_j \sum_l \beta_j \gamma_{jl} \varrho_l = \frac{1}{6} \tag{7.37}$$

are fulfilled; and finally, the order of consistency $p = 4$ iff in addition the four conditions

$$\sum_j \beta_j \varrho_j^3 = \frac{1}{4}, \qquad \sum_j \sum_l \beta_j \varrho_j \gamma_{jl} \varrho_l = \frac{1}{8},$$
$$\sum_j \sum_l \beta_j \gamma_{jl} \varrho_l^2 = \frac{1}{12}, \qquad \sum_j \sum_l \sum_r \beta_j \gamma_{jl} \gamma_{lr} \varrho_r = \frac{1}{24} \tag{7.38}$$

are satisfied. In each summation the index runs from 1 to m.

Proof. See, e.g., [19]. ∎

Obviously, the number of the order conditions which have to be fulfilled by the coefficients of a Runge-Kutta method rises with growing order. Table 7.2 gives an impression of that fact.

In the second step, the order conditions must be solved. To simplify the representation, we want to assume that the coefficients of an (explicit) ERK should be determined. Thus, $4(4 + 1)/2 = 10$ coefficients β, Γ of a 4-stage RKM are searched which satisfy the 8 nonlinear algebraic equations (7.35)–(7.38). More precisely, this under-determined system reads

$$\beta_1 + \beta_2 + \beta_3 + \beta_4 = 1, \quad \beta_2 \varrho_2 + \beta_3 \varrho_3 + \beta_4 \varrho_4 = \frac{1}{2}, \tag{7.39}$$

$$\beta_2 \varrho_2^2 + \beta_3 \varrho_3^2 + \beta_4 \varrho_4^2 = \frac{1}{3}, \tag{7.40a}$$

$$\beta_3 \gamma_{32} \varrho_2 + \beta_4 (\gamma_{42} \varrho_2 + \gamma_{43} \varrho_3) = \frac{1}{6}, \tag{7.40b}$$

$$\beta_2 \varrho_2^3 + \beta_3 \varrho_3^3 + \beta_4 \varrho_4^3 = \frac{1}{4}, \tag{7.41a}$$

$$\beta_3 \varrho_3 \gamma_{32} \varrho_2 + \beta_4 \varrho_4 (\gamma_{42} \varrho_2 + \gamma_{43} \varrho_3) = \frac{1}{8}, \tag{7.41b}$$

$$\beta_3 \gamma_{32} \varrho_2^2 + \beta_4 (\gamma_{42} \varrho_2^2 + \gamma_{43} \varrho_3^2) = \frac{1}{12}, \tag{7.42a}$$

$$\beta_4 \gamma_{43} \gamma_{32} \varrho_2 = \frac{1}{24}. \tag{7.42b}$$

In addition, the relations (7.25) must be valid:

$$\varrho_1 = 0, \tag{7.43a}$$

$$\varrho_2 = \gamma_{21}, \tag{7.43b}$$

$$\varrho_3 = \gamma_{31} + \gamma_{32}, \tag{7.43c}$$

$$\varrho_4 = \gamma_{41} + \gamma_{42} + \gamma_{43}. \tag{7.43d}$$

To determine special sets of coefficients from the above equations, we want to consider the formulae (7.39)–(7.40a) and (7.41a) from the point of view of a pure integration. Let us assume that the coefficients β_1, β_2, β_3, and β_4 are the weights and the coefficients $\varrho_1 = 0$, ϱ_2, ϱ_3, and ϱ_4 are the nodes of a quadrature formula on the interval $[0, 1]$,

$$\int_0^1 f(x)dx \approx \sum_{j=1}^4 \beta_j f(\varrho_j), \tag{7.44}$$

which integrates polynomials of degree 3 exactly. As is generally known, Simpson's rule,

$$\int_0^1 f(t)\, dt \approx \frac{1}{6} (f(0) + 4f(1/2) + f(1))$$

$$= \frac{1}{6} (f(0) + 2f(1/2) + 2f(1/2) + f(1)),$$

Fig. 7.13 Newton's 3/8 rule

$$
\begin{array}{c|cccc}
0 & & & & \\
\dfrac{1}{3} & \dfrac{1}{3} & & & \\
\dfrac{2}{3} & -\dfrac{1}{3} & 1 & & \\
1 & 1 & -1 & 1 & \\
\hline
 & \dfrac{1}{8} & \dfrac{3}{8} & \dfrac{3}{8} & \dfrac{1}{8}
\end{array}
$$

is such a formula (see, e.g., [20]). Since this equation is based on three nodes only, we have used the middle node twice and obtain for the right-hand side of (7.44) the following coefficients:

$$
\varrho = (0, \frac{1}{2}, \frac{1}{2}, 1)^T, \qquad \beta = (\frac{1}{6}, \frac{2}{6}, \frac{2}{6}, \frac{1}{6})^T.
$$

Now Eq. (7.43b) yields $\gamma_{21} = 1/2$. With $z \equiv (\gamma_{42}\varrho_2 + \gamma_{43}\varrho_3)$, Eqs. (7.40b) and (7.41b) read

$$
\gamma_{32} + z = 1 \quad \text{und} \quad 2\gamma_{32} + 4z = 3,
$$

and it follows $\gamma_{32} = 1/2$ and $z = 1/2$. From (7.43c), we get $\gamma_{31} = 0$. Furthermore, (7.42b) implies $\gamma_{43} = 1$. It is $z = 1/2 = (1/2)\gamma_{42} + (1/2)$, i.e., $\gamma_{42} = 0$. Finally, from Eq. (7.43d), we obtain $\gamma_{41} = 0$. Looking at Fig. 7.6, it can be easily recognized that we have determined the coefficients of the classical Runge-Kutta method.

Another quadrature formula with the above stated property is Newton's 3/8 rule:

$$
\int_0^1 f(x)\,dx \approx \frac{1}{8}\left(f(0) + 3f(1/3) + 3f(2/3) + f(1)\right).
$$

Here, we choose

$$
\varrho = (0, \frac{1}{3}, \frac{2}{3}, 1)^T, \qquad \beta = (\frac{1}{8}, \frac{3}{8}, \frac{3}{8}, \frac{1}{8})^T.
$$

An analogous calculation leads to the following result:

$$
\gamma_{21} = \frac{1}{3}, \; \gamma_{32} = 1, \; z = \frac{1}{3}, \; \gamma_{31} = -\frac{1}{3}, \; \gamma_{43} = 1, \; \gamma_{42} = -1, \; \gamma_{41} = 1.
$$

This Runge-Kutta method is called Newton's 3/8 rule, too. The associated Butcher diagram is given in Fig. 7.13.

7.6 Collocation and Implicit Runge-Kutta Methods

We start with the following definition.

Definition 7.20. Let $\varrho_1, \ldots, \varrho_m$ be mutually distinct real numbers with $0 \leq \varrho_j \leq 1$, $j = 1, \ldots, m$. The collocation polynomial $P_m(x)$ is a polynomial of maximum degree m which satisfies the collocation conditions

$$P_m(x_i) = y_i \tag{7.45}$$

$$\dot{P}_m(x_i + \varrho_j h) = f(x_i + \varrho_j h, P_m(x_i + \varrho_j h)), \quad j = 1, \ldots, m. \tag{7.46}$$

The *numerical solution* of the collocation method is defined by

$$y_{i+1} \equiv P_m(x_i + h). \tag{7.47}$$

For $m = 1$, the collocation method has the form

$$P_m(x) = y_i + (x - x_i)k, \quad \text{with} \quad k \equiv f(x_i + \varrho_1 h, y_i + h\varrho_1 k).$$

Thus, Euler's (forward) method with $\varrho_1 = 0$, Euler's (backward) method with $\varrho_1 = 1$ and the midpoint rule with $\varrho_1 = 1/2$ are collocation methods.

The close connection between the collocation methods and the Runge-Kutta methods is established by the following theorem.

Theorem 7.21. *The collocation method is equivalent to an m-stage implicit RKM with the coefficients*

$$\gamma_{jl} \equiv \int_0^{\varrho_j} L_l(\tau) d\tau, \quad \beta_j \equiv \int_0^1 L_j(\tau) d\tau, \quad j, l = 1, \ldots, m, \tag{7.48}$$

where $L_j(\tau)$ denotes the jth Lagrange factor

$$L_j(\tau) = \prod_{l=1, l \neq j}^{m} \frac{\tau - \varrho_l}{\varrho_j - \varrho_l}.$$

Proof. See, e.g., [16].   ∎

Remark 7.22. Since $\tau^{k-1} = \sum_{j=1}^{m} \varrho_j^{k-1} L_j(\tau)$, $k = 1, \ldots, m$, Eqs. (7.48) are equivalent to the linear systems

$$C(q): \quad \sum_{l=1}^{m} \gamma_{jl} \varrho_l^{k-1} = \frac{\varrho_j^k}{k}, \quad k = 1, \ldots, q, \; j = 1, \ldots, m,$$

$$B(p): \quad \sum_{j=1}^{m} \beta_j \varrho_j^{k-1} = \frac{1}{k}, \quad k = 1, \ldots, p, \tag{7.49}$$

with $q = m$ and $p = m$.

Equations (7.49) are part of the so-called simplifying assumptions which have been introduced in [6]. The following equations still belong to these assumptions:

$$D(r): \quad \sum_{j=1}^{m} \beta_j \varrho_j^{k-1} \gamma_{jl} = \frac{1}{k} \beta_l (1 - \varrho_l^k), \quad l = 1, \ldots, m, \quad k = 1, \ldots, r.$$

$$\tag{7.50}$$

The simplifying assumptions make it possible to develop the Runge-Kutta methods with a high order of consistency. Namely, if these equations are fulfilled, the number of order conditions decreases which have to be satisfied to guarantee a prescribed order of consistency.

The simplifying assumptions permit a simple interpretation. Let us consider the special IVP

$$\dot{y}(x) = f(x), \quad y(x_i) = 0, \tag{7.51}$$

with the exact solution:

$$y(x_i + h) = h \int_0^1 f(x_i + \theta h) \, d\theta.$$

If an m-stage Runge-Kutta method is applied to (7.51), we get

$$y_{i+1}^h = h \sum_{j=1}^{m} \beta_j f(x_i + \varrho_j h).$$

In the case $f(x) = (x - x_i)^{k-1}$, the exact solution is

$$y(x_i + h) = h \int_0^1 (\theta h)^{k-1} \, d\theta = \frac{1}{k} h^k$$

and the numerical solution is

$$y_{i+1}^h = h^k \sum_{j=1}^{m} \beta_j \varrho_j^{k-1}.$$

Then, the condition $B(p)$ implies that the quadrature formula on which the Runge-Kutta method is based integrates exactly the integrals

$$\int_0^1 \theta^{k-1} d\theta, \quad k = 1, \ldots, p.$$

In other words, polynomials up to the degree $p - 1$ are exactly integrated on the interval $[0, 1]$, i.e., the quadrature formula has the order of accuracy $p - 1$.

From the simplifying assumption, $C(q)$ follows accordingly that the intermediary values y_{ij}^h, $j = 1, \ldots, m$, which occur in a Runge-Kutta method (see formula (7.14)), are determined by a quadrature formula of the order of accuracy $q - 1$. In particular, the simplifying assumption $C(1)$ yields the condition (7.25) on the nodes of a Runge-Kutta method.   □

The fundamental relation between the RKMs and the collocation methods is formulated in the following theorem.

Theorem 7.23. *An m-stage IRK with mutually distinct nodes $\varrho_1, \ldots, \varrho_m$ and an order of consistency $p \geq m$ is a collocation method iff the simplifying assumption $C(m)$ is true.*

Proof. See the monograph [16].   ■

About the order of consistency of an IRK which is determined by the coefficients (7.48), the following can be stated.

Theorem 7.24. *If the simplifying assumption $B(p)$ is satisfied for a $p \geq m$, then the collocation method has the order of consistency p. This means that the collocation method has the same order of consistency as the underlying quadrature formula.*

Proof. See the monograph [26].   ■

Theorem 7.24 shows that the choice of an appropriate quadrature formula is of great importance for the construction of IRKs. A class of IRKs is based on the theory of the Gauss-Legendre quadrature formulae (see, e.g., [20]). Therefore, these IRKs are called Gauss methods. The nodes $\varrho_1, \ldots, \varrho_m$ of a Gauss method are the mutually distinct roots of the shifted Legendre polynomial of degree m

$$\hat{\phi}_m(t) \equiv \phi_m(2t - 1) = \frac{1}{m!} \frac{d^m}{d\,t^m} \left(t^m (t - 1)^m \right),$$

i.e., they are the Gauss-Legendre points in the interval $(0, 1)$. Tables of these roots can be found in the corresponding handbooks of mathematical functions like [1]. Because the quadrature formula used in this context has the order of accuracy $2m$, Theorem 7.24 implies that an IRK (more precisely, a collocation method) which is based on these nodes has also the order $p = 2m$.

For $m = 1$ immediately results the (implicit) midpoint rule with the order of consistency $p = 2m = 2$. The coefficients of the IRKs with the stage numbers $m = 2$ and $m = 3$ are given in Figs. 7.14 and 7.15, respectively.

Fig. 7.14 Gauss method of order 4

$$
\begin{array}{c|cc}
\dfrac{1}{2} - \dfrac{\sqrt{3}}{6} & \dfrac{1}{4} & \dfrac{1}{4} - \dfrac{\sqrt{3}}{6} \\[2ex]
\dfrac{1}{2} + \dfrac{\sqrt{3}}{6} & \dfrac{1}{4} + \dfrac{\sqrt{3}}{6} & \dfrac{1}{4} \\[2ex]
\hline
& \dfrac{1}{2} & \dfrac{1}{2}
\end{array}
$$

$$
\begin{array}{c|ccc}
\dfrac{1}{2} - \dfrac{\sqrt{15}}{10} & \dfrac{5}{36} & \dfrac{2}{9} - \dfrac{\sqrt{15}}{15} & \dfrac{5}{36} - \dfrac{\sqrt{15}}{30} \\[2ex]
\dfrac{1}{2} & \dfrac{5}{36} + \dfrac{\sqrt{15}}{24} & \dfrac{2}{9} & \dfrac{5}{36} - \dfrac{\sqrt{15}}{24} \\[2ex]
\dfrac{1}{2} + \dfrac{\sqrt{15}}{10} & \dfrac{5}{36} + \dfrac{\sqrt{15}}{30} & \dfrac{2}{9} + \dfrac{\sqrt{15}}{15} & \dfrac{5}{36} \\[2ex]
\hline
& \dfrac{5}{18} & \dfrac{4}{9} & \dfrac{5}{18}
\end{array}
$$

Fig. 7.15 Gauss method of order 6

For a given stage number m, the Gauss methods have the highest possible order of an IRK. However, since their stability properties are not optimum (see Sect. 7.9), there are IRKs with an order of consistency $p = 2m - 1$ whose stability is better. In particular, such methods are used in the numerical treatment of so-called *stiff* equations.

The basis for the construction of these IRKs supplies the following theorem.

Theorem 7.25. *Let an RKM have the order of consistency* $p = 2m - 1$. *Then, the nodes* $\varrho_1, \ldots, \varrho_m$ *are the roots of a polynomial*

$$\phi_{m,\xi}(2t - 1) = \phi_m(2t - 1) + \xi\,\phi_{m-1}(2t - 1), \quad \xi \in \mathbb{R}.$$

Proof. See, e.g., [25]. ∎

Besides, the cases $\xi = 1$ and $\xi = -1$ are especially interesting. The accompanying quadrature formulae are the left-sided ($\varrho_1 = 0$) and the right-sided ($\varrho_m = 1$) Radau quadrature methods. Both have the order of accuracy $2m - 1$. The resulting IRKs are called *Radau I methods* and *Radau II methods*, respectively. The nodes of a Radau method are mutually distinct and lie in the interval $[0, 1)$. More specifically, it holds:

- The nodes $\varrho_1 = 0$, $\varrho_2, \ldots, \varrho_m$ of a Radau I method are the roots of the polynomial

$$\phi_1(t) \equiv \frac{d^{m-1}}{d\,t^{m-1}}\big(t^m(t-1)^{m-1}\big).$$

Fig. 7.16 Radau IA methods
of order 1 and 3

$$
\begin{array}{c|cc}
0 & \dfrac{1}{4} & -\dfrac{1}{4} \\[2mm]
\dfrac{2}{3} & \dfrac{1}{4} & \dfrac{5}{12} \\[2mm]
\hline
 & \dfrac{1}{4} & \dfrac{3}{4}
\end{array}
$$

$$
\begin{array}{c|c}
0 & 1 \\
\hline
 & 1
\end{array}
$$

Fig. 7.17 Radau IA method
of order 5

$$
\begin{array}{c|ccc}
0 & \dfrac{1}{9} & \dfrac{-1-\sqrt{6}}{18} & \dfrac{-1+\sqrt{6}}{18} \\[3mm]
\dfrac{6-\sqrt{6}}{10} & \dfrac{1}{9} & \dfrac{88+7\sqrt{6}}{360} & \dfrac{88-43\sqrt{6}}{360} \\[3mm]
\dfrac{6+\sqrt{6}}{10} & \dfrac{1}{9} & \dfrac{88+43\sqrt{6}}{360} & \dfrac{88-7\sqrt{6}}{360} \\[3mm]
\hline
 & \dfrac{1}{9} & \dfrac{16+\sqrt{6}}{36} & \dfrac{16-\sqrt{6}}{36}
\end{array}
$$

- The nodes $\varrho_1,\ldots,\varrho_{m-1},\varrho_m = 1$ of a Radau II method are the roots of the polynomial

$$
\phi_2(t) \equiv \frac{d^{m-1}}{d\,t^{m-1}}\left(t^{m-1}(t-1)^m\right).
$$

The coefficients β_j of a Radau method are determined by the simplifying assumption $B(m)$ (see formula (7.49)). For the choice of the elements γ_{jk} of the matrix Γ, there are different proposals in the literature:

- Radau I method: Γ is determined by the simplifying assumption $C(m)$ (see [4]).
- Radau IA method: Γ is determined by the simplifying assumption $D(m)$ (see [11]).
- Radau II method: Γ is determined by the simplifying assumption $D(m)$ (see [4]).
- Radau IIA method: Γ is determined by the simplifying assumption $C(m)$ (see [11]).

Regarding the stability behavior, the Radau IA and Radau IIA methods mentioned in the above listing are improved variants of the Radau I and Radau II methods, respectively. Therefore, they are used in practice preferentially. Finally, the order of accuracy of the used quadrature formula implies that the m-stage Radau I, Radau IA, Radau II, and Radau IIA methods have the order of consistency $p = 2m - 1$.

In Figs. 7.16 and 7.17, the coefficients of the Radau IA methods with $m = 1, 2, 3$ are given.

Fig. 7.18 Radau IIA
methods of order 1 and 3

$$
\begin{array}{c|c}
1 & 1 \\
\hline
 & 1
\end{array}
\qquad
\begin{array}{c|cc}
\dfrac{1}{3} & \dfrac{5}{12} & -\dfrac{1}{12} \\[2mm]
1 & \dfrac{3}{4} & \dfrac{1}{4} \\[2mm]
\hline
 & \dfrac{3}{4} & \dfrac{1}{4}
\end{array}
$$

$$
\begin{array}{c|ccc}
\dfrac{4-\sqrt{6}}{10} & \dfrac{88-7\sqrt{6}}{360} & \dfrac{296-169\sqrt{6}}{1800} & \dfrac{-2+3\sqrt{6}}{225} \\[3mm]
\dfrac{4+\sqrt{6}}{10} & \dfrac{296+169\sqrt{6}}{1800} & \dfrac{88+7\sqrt{6}}{360} & \dfrac{-2-3\sqrt{6}}{225} \\[3mm]
1 & \dfrac{16-\sqrt{6}}{36} & \dfrac{16+\sqrt{6}}{36} & \dfrac{1}{9} \\[3mm]
\hline
 & \dfrac{16-\sqrt{6}}{36} & \dfrac{16+\sqrt{6}}{36} & \dfrac{1}{9}
\end{array}
$$

Fig. 7.19 Radau IIA method of order 5

In the Figs. 7.18 and 7.19, the coefficients of the Radau IIA methods with $m =$ 1, 2, 3 are given.

A further class of RKMs, which play an important role for the applications, are of the order of consistency $p = 2m - 2$. They are based on the following statement.

Theorem 7.26. *Let an RKM have the order of consistency $p = 2m - 2$. Then, the nodes $\varrho_1, \ldots, \varrho_m$ are the roots of a polynomial*

$$
\phi_{m,\xi,\mu}(2t - 1) \equiv \phi_m(2t - 1) + \xi\phi_{m-1}(2t - 1) + \mu\phi_{m-2}(2t - 1), \quad \xi, \mu \in \mathbb{R}.
$$

Proof. See, e.g., [25].   ■

The really interesting case is $\xi = 0$ and $\mu = -1$ because from this choice, the so-called Lobatto methods arise. Lobatto quadrature formulae have the greatest possible order of accuracy iff the endpoints $\varrho_1 = 0$ and $\varrho_m = 1$ are included in the set of nodes. For historical reasons, the resulting RKMs are called *Lobatto III methods*.

The nodes of a Lobatto method are mutually distinct and lie in the closed interval [0, 1]. More specifically, it holds:

- The nodes $\varrho_1 = 0, \varrho_2, \ldots, \varrho_{m-1}, \varrho_m = 1$ of a Lobatto III method are the roots of the polynomial

$$
\phi_3(t) \equiv \frac{d^{m-2}}{dt^{m-2}}\left(t^{m-1}(t-1)^{m-1}\right).
$$

Fig. 7.20 Lobatto IIIA
method of order 4

$$
\begin{array}{c|ccc}
0 & 0 & 0 & 0 \\
\dfrac{1}{2} & \dfrac{5}{24} & \dfrac{1}{3} & -\dfrac{1}{24} \\
1 & \dfrac{1}{6} & \dfrac{2}{3} & \dfrac{1}{6} \\
\hline
 & \dfrac{1}{6} & \dfrac{2}{3} & \dfrac{1}{6}
\end{array}
$$

$$
\begin{array}{c|cccc}
0 & 0 & 0 & 0 & 0 \\
\dfrac{5-\sqrt{5}}{10} & \dfrac{11+\sqrt{5}}{120} & \dfrac{25-\sqrt{5}}{120} & \dfrac{25-13\sqrt{5}}{120} & \dfrac{-1+\sqrt{5}}{120} \\
\dfrac{5+\sqrt{5}}{10} & \dfrac{11-\sqrt{5}}{120} & \dfrac{25+13\sqrt{5}}{120} & \dfrac{25+\sqrt{5}}{120} & \dfrac{-1-\sqrt{5}}{120} \\
1 & \dfrac{1}{12} & \dfrac{5}{12} & \dfrac{5}{12} & \dfrac{1}{12} \\
\hline
 & \dfrac{1}{12} & \dfrac{5}{12} & \dfrac{5}{12} & \dfrac{1}{12}
\end{array}
$$

Fig. 7.21 Lobatto IIIA method of order 6

The coefficients β_j are determined by the underlying quadrature formula. For the choice of the elements γ_{jk} of the matrix Γ, there are different proposals in the literature:

- Lobatto IIIA methods: Γ is determined by the simplifying assumption $C(m)$ (see [11]).
- Lobatto IIIB methods: Γ is determined by the simplifying assumption $D(m)$ (see [11]).
- Lobatto IIIC methods: Γ is determined by the simplifying assumption $C(m-1)$ and the additional conditions $\gamma_{j1} = \beta_1$, $j = 1,\ldots,m$ (see [7]).

The order of accuracy of the used quadrature formula implies that the m-stage Lobatto IIIA, Lobatto IIIB, and Lobatto IIIC methods have the order of consistency $p = 2m - 2$.

In Figs. 7.20 and 7.21, the coefficients of the Lobatto IIIA methods with $m = 3$ and $m = 4$ are given.

Finally, in Figs. 7.22 and 7.23, we present the coefficients of the Lobatto IIIB methods with $m = 3$ and $m = 4$.

In summary it remains to be noticed that according to the statement of Theorem 7.23, the Gauss methods, the Radau I methods, the Radau IIA methods, and the Lobatto IIIA methods belong to the class of the collocation methods.

Fig. 7.22 Lobatto IIIB
method of order 4

$$
\begin{array}{c|ccc}
0 & \dfrac{1}{6} & -\dfrac{1}{6} & 0 \\[2mm]
\dfrac{1}{2} & \dfrac{1}{6} & \dfrac{1}{3} & 0 \\[2mm]
1 & \dfrac{1}{6} & \dfrac{5}{6} & 0 \\[2mm]
\hline
 & \dfrac{1}{6} & \dfrac{2}{3} & \dfrac{1}{6}
\end{array}
$$

$$
\begin{array}{c|cccc}
0 & \dfrac{1}{12} & \dfrac{-1-\sqrt5}{24} & \dfrac{-1+\sqrt5}{24} & 0 \\[3mm]
\dfrac{5-\sqrt5}{10} & \dfrac{1}{12} & \dfrac{25+\sqrt5}{120} & \dfrac{25-13\sqrt5}{120} & 0 \\[3mm]
\dfrac{5+\sqrt5}{10} & \dfrac{1}{12} & \dfrac{25+13\sqrt5}{120} & \dfrac{25-\sqrt5}{120} & 0 \\[3mm]
1 & \dfrac{1}{12} & \dfrac{11-\sqrt5}{24} & \dfrac{11+\sqrt5}{24} & 0 \\[3mm]
\hline
 & \dfrac{1}{12} & \dfrac{5}{12} & \dfrac{5}{12} & \dfrac{1}{12}
\end{array}
$$

Fig. 7.23 Lobatto IIIB method of order 6

7.7 Global Error and Convergence

The study of the local discretization error presupposes that the integration of the
IVP (7.5) is realized over a single interval $[x_i, x_{i+1}]$, where at the left boundary x_i
the exact value of the solution $y(x)$ is given. But obviously that is not what happens
in reality. Instead, the integration is done successively on all intervals $[x_j, x_{j+1}]$,
$j = 0,\ldots,N-1$, which are defined by the underlying grid J_h. Already at the
end of the first interval $[x_0, x_1]$, we have only an approximation y_1^h for $y(x_1)$. The
important point is that in the approximation y_{i+1}^h, all previous local errors have
gradually added up. This accumulation of the local errors is measured by the *global
error*.

Definition 7.27. As before, let y_{i+1}^h be the approximation of the exact solution
$y(x_{i+1})$ which has been determined by a one-step method with the step-size h. The
global error of this approximation is defined as

$$
e_{i+1}^h \equiv y(x_{i+1}) - y_{i+1}^h. \tag{7.52}
$$

$\square$

An important information on the global error is formulated in the following theorem.

Theorem 7.28. *Assume that* $f \in \mathrm{Lip}(S)$ *(see formula (7.6)) and* L *denotes the corresponding Lipschitz constant. Let the upper integration limit* $T (< \infty)$ *be chosen such that the solution trajectory of the IVP remains in the closed region* Ω. *Then, there exists a constant* C *such that for all* i *with* $x_0 + (i + 1)h \le T$, *the following estimation holds:*

$$\|e_{i+1}^h\| \le C \max_{j \le i} \|\delta(x_{j+1}, y_j(x_{j+1}); h)\|. \tag{7.53}$$

In particular, we have $\|e_{i+1}^h\| = O(h^p)$, *where* p *denotes the order of consistency of the used one-step method.*

Proof. See, e.g., [18]. ∎

Now, we consider the convergence of a one-step method which describes the fact that the *exact* solution of the difference equation (one-step method) converges to the *exact* solution of the IVP for $h \to 0$.

Definition 7.29. A one-step method is called convergent if there exists a region Ω (see Theorem 7.28) such that for a fixed value $x \equiv x_0 + (i + 1)h \le T$, it holds

$$\lim_{h \to 0, i \to \infty} e_{i+1}^h = 0. \tag{7.54}$$

The largest positive integer q for which $e_{i+1}^h = O(h^q)$ is called order of convergence. □

Theorem 7.28 implies the important statement that a one-step method has the order of convergence p if and only if it is consistent and has the order of consistency p. Since in that case the order of convergence is identical with the order of consistency, we will use the uniform designation *order* for a one-step method.

7.8 Estimation of the Local Discretization Error and Step-Size Control

In the previous sections, we have assumed that the step-size h is fixed. Let us now consider a *nonuniform grid*, and we want to adjust the step-size to the growth behavior of the exact solution $y(x)$ of the given IVP. To be able to develop an appropriate step-size control, a sufficiently accurate estimation of the local discretization error is required. Moreover, this estimation must be easy to implement without a large additional effort. Today one of the most frequently used techniques to estimate the local discretization error is based on embedded Runge-Kutta methods (see Sect. 7.3, Figs. 7.7–7.9). Therefore, it is called *estimation by the embedding principle*. However, this technique can also be realized more generally by two one-step methods with different orders.

Let us assume that two explicit one-step methods are given:

$$y_{i+1}^h = y_i^h + h \, \Phi_1(x_i, y_i^h; h) \quad \text{and} \quad y_{i+1}^h = y_i^h + h \, \Phi_2(x_i, y_i^h; h), \qquad (7.55)$$

where

$$y_{i+1}^h = y_i(x_{i+1}) - h\delta(x_{i+1}, y_i(x_{i+1}); h), \quad \delta(\cdot) = O(h^p),$$
$$\bar{y}_{i+1}^h = y_i(x_{i+1}) - h\bar{\delta}(x_{i+1}, y_i(x_{i+1}); h), \quad \bar{\delta}(\cdot) = O(h^q), \qquad (7.56)$$

and $q \geq p + 1$. Subtracting the first equation from the second one gives

$$\bar{y}_{i+1}^h - y_{i+1}^h = h\,\delta(\cdot) - h\,\bar{\delta}(\cdot).$$

Thus,

$$\frac{1}{h}\left(\bar{y}_{i+1}^h - y_{i+1}^h\right) + \underbrace{\bar{\delta}(\cdot)}_{O(h^q)} = \underbrace{\delta(\cdot)}_{O(h^p)}.$$

From this follows immediately that an appropriate estimate for the local discretization error can be defined as

$$s \equiv \frac{1}{h}\left(\bar{y}_{i+1}^h - y_{i+1}^h\right). \qquad (7.57)$$

If a pair of adjacent methods from a class of embedded Runge-Kutta methods is used for this estimation technique, then it is not necessary to compute two different methods independently of one another. For example, the use of the methods RKF4 and RKF5 (see Fig. 7.7) with $p = 4$ and $q = 5$ leads to the following estimate which can be determined with a minimum additional computational work:

$$\delta(x_{i+1}, y_i(x_{i+1}); h) \approx s = \sum_{j=1}^{6}(\bar{\beta}_j - \beta_j)k_j, \qquad (7.58)$$

where $\{\beta_j\}_{j=1}^5$ ($\beta_6 = 0$) are the parameters of RKF4 and $\{\bar{\beta}_j\}_{j=1}^6$ are the parameters of RKF5.

Another strategy to estimate the local discretization error is the so-called Runge principle. Here, only a single one-step method is used. But the integration from x_i to x_{i+1} is executed twice, namely, with two different step-sizes. If these step-sizes are h and $h/2$, the following estimate for the local discretization error results (see, e.g., [18]):

$$s \equiv \frac{1}{\left(1 - \left(\frac{1}{2}\right)^p\right)h}\left(\bar{y}_{i+1}^h - y_{i+1}^h\right), \qquad (7.59)$$

where y_{i+1}^h is the approximation of $y(x_{i+1})$ which has been computed with the step-size h and $\bar{y}_{i+1}^h$ is the approximation obtained with the step-size $h/2$.

Now we want to show how the step-size can be controlled. Let us simplify the representation by restricting to *explicit* one-step methods only. The local step-sizes are denoted by $h_i \equiv x_{i+1} - x_i$. For a given (absolute) tolerance TOL on the norm of the local discretization error, the following two criteria are normally used:

- *Error per step* (EPS): the step-size h_i is chosen such that

$$\|h_i \, \delta(x_{i+1}, y_i^{h_i}(x_{i+1}); h_i)\| \approx \text{TOL}, \tag{7.60}$$

- *Error per unit step* (EPUS): the step-size h_i is chosen such that

$$\|\delta(x_{i+1}, y_i^{h_i}(x_{i+1}); h_i)\| \approx \text{TOL}. \tag{7.61}$$

For the realization of the EPS criterion, let us assume that an estimate EST for $\|h_i \, \delta(x_{i+1}, y_i^{h_i}(x_{i+1}); h_i)\|$ is known, with EST $\neq 0$. Furthermore, let α be a real constant which is defined by

$$\alpha \equiv 0.9 \left(\frac{\text{TOL}}{\text{EST}}\right)^{\frac{1}{p+1}}. \tag{7.62}$$

Now, the step-size is controlled by the following strategy:

- If EST/TOL ≤ 1, the actual step-size h_i will be accepted and the step-size h_{i+1} of the next integration step on the interval $[x_{i+1}, x_{i+1} + h_{i+1}]$ is increased by the formula

$$h_{i+1} = \alpha \, h_i. \tag{7.63}$$

- If EST/TOL > 1, the actual step-size $h_i^{\text{old}} \equiv h_i$ will not be accepted. Instead, a new step-size is determined by the formula

$$h_i^{\text{new}} = \alpha \, h_i^{\text{old}}. \tag{7.64}$$

Then, the one-step method is applied on the reduced interval $[x_i, x_i + h_i^{\text{new}}]$, and at the end of this interval, the local discretization error of the approximation is estimated.

The formulae (7.63) and (7.64) are based on the fact that $\|h \, \delta(\cdot)\| = O(h^{p+1})$, i.e., $h \equiv h_{i+1}$ and $h \equiv h_i^{\text{new}}$ are determined such that the relation

$$\left[\frac{h}{h_i}\right]^{p+1} \approx \frac{\text{TOL}}{\text{EST}} \tag{7.65}$$

is fulfilled. In practice, the use of a so-called safety factor in (7.62) has proven to be favorable. Here, we have used the value 0.9. In some implementations the value 0.8 can be seen instead.

If EST $\approx$ 0, the constant α cannot be computed by (7.62). Therefore, in the implementations of this control strategy, the user must prescribe a maximum step-size h_{max} and a minimum step-size h_{min} which are used in that exceptional situation.

In some computer codes, α is not always determined by (7.62). For instance, the IVP-codes in MATLAB proceed as follows. If the step-size has been reduced by (7.62) and (7.64), and at the end of the new interval the test EST/TOL $\leq$ 1 is not fulfilled, $\alpha = 0.5$ is used in (7.64) until this inequality is satisfied.

If the EPUS criterion is used instead of the EPS criterion, EST is an estimate for $\|\delta(x_{i+1}, y_i^{h_i}; h_i)\|$. Moreover, in the formulae (7.62) and (7.65), the term $p + 1$ has to be replaced by p.

The error control may also be performed in a *relative* way:

- *Relative error per step* (REPS): the step-size h_i is chosen such that

$$\frac{\|h_i \, \delta(x_{i+1}, y_i^{h_i}(x_{i+1}); h_i)\|}{\|y_{i+1}^{h_i}\|} \approx \text{TOL}, \qquad (7.66)$$

- *Relative error per unit step* (REPUS): the step-size h_i is chosen such that

$$\frac{\|\delta(x_{i+1}, y_i^{h_i}(x_{i+1}); h_i)\|}{\|y_{i+1}^{h_i}\|} \approx \text{TOL}. \qquad (7.67)$$

To take the error control as efficiently and effectively as possible, a combination of the EPS criterion and the REPS criterion (EPUS criterion and REPUS criterion, respectively) is often used. Then, in addition to the absolute tolerance TOL, a relative tolerance RTOL must be supplied by the user. The step-size is determined such that the following relation holds:

$$\text{EST} \approx \text{TOL} + \text{RTOL} \, \|y_{i+1}^{h_i}\|. \qquad (7.68)$$

Example 7.30. The influence of the order of consistency of an RKM and the step-size control used on the accuracy of computation can be demonstrated by the well-known predator-prey relationship:

$$y_1'(x) = y_1(x) \, (\varepsilon_1 - \gamma_1 y_2(x)) ,$$
$$y_2'(x) = -y_2(x) \, (\varepsilon_2 - \gamma_2 y_1(x)) ,$$
$$y_1(0) = 700, \quad y_2(0) = 500 \qquad (7.69)$$

(see, e.g., http://en.wikipedia.org/wiki/Lotka-Volterra_equation).

Here, y_1 denotes the number of prey and y_2 the number of predators. For special choices of the parameters ε_i and γ_i, $i = 1, 2$, the IVP (7.69) possesses a

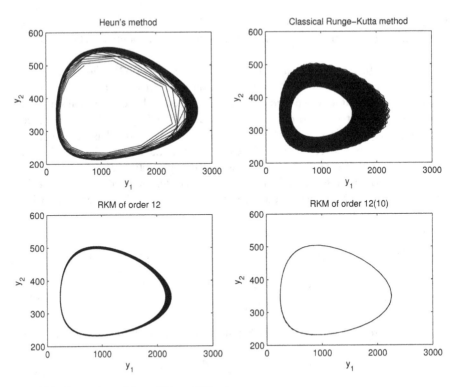

Fig. 7.24 Comparison of three different IVP-solvers

solution which changes periodically in time. For our test computations, we have used $\varepsilon_1 = 0.7$, $\varepsilon_2 = 0.09$, $\gamma_1 = 0.002$, and $\gamma_2 = 0.0001$. In Fig. 7.24, the results are presented which have been obtained with four different ERKs (Heun's method, classical Runge-Kutta method, RKM of order 12, and RKM of order 12(10); for the last two methods, see [22]). The integration interval is $[0, 10,000]$, and y_2 versus y_1 is shown in each plot. The first three methods have been realized with the same constant step-size $h = 0.2$. In the fourth method, the step-size is automatically controlled by the local discretization error (we have used the tolerance TOL $= 1e - 7$). Obviously, in the fourth plot the exact solution trajectory is quite well reproduced, whereas the first three plots show that the local errors are accumulated such that no periodic solution results.   □

7.9   Absolute Stability and Stiffness

The properties of one-step methods hitherto discussed, like consistency and convergence, are asymptotic concepts, i.e., they are based on the assumption that $h \to 0$. However, if the method is realized on a computer, the step-size is very small but different from zero. Therefore, indicators are of interest which characterize the behavior of a one-step method for step-sizes $h \neq 0$. Such an indicator is the *absolute stability*. The basis of this stability concept is the following test problem which has been introduced by Dahlquist in 1963 (see [8]):

$$y'(x) = \lambda \, y(x), \quad y(0) = 1, \tag{7.70}$$

where $\lambda \in \mathbb{R}$ or $\lambda \in \mathbb{C}$. The exact solution of this IVP is $y(x) = e^{\lambda x}$.

At first we want to study the behavior of the classical Runge-Kutta method (7.17) when it is applied to the test problem. Since the corresponding ODE is autonomous, the difference equations of the classical RKM can be simplified:

$$y_{i+1}^h = y_i^h + h \left(\frac{1}{6} k_1 + \frac{1}{3} k_2 + \frac{1}{3} k_3 + \frac{1}{6} k_4 \right),$$

with

$$k_1 = f(y_i^h), \quad k_2 = f\left(y_i^h + \frac{1}{2} h \, k_1\right), \quad k_3 = f\left(y_i^h + \frac{1}{2} h \, k_2\right),$$

$$k_4 = f(y_i^h + h \, k_3).$$

For the IVP (7.70), these equations read

$$k_1 = \lambda y_i^h,$$

$$k_2 = \lambda \left(y_i^h + \frac{1}{2} h k_1\right) = \left(\lambda + \frac{1}{2} h \lambda^2\right) y_i^h,$$

$$k_3 = \lambda \left(y_i^h + \frac{1}{2} h k_2\right) = \left(\lambda + \frac{1}{2} h \lambda^2 + \frac{1}{4} h^2 \lambda^3\right) y_i^h,$$

$$k_4 = \lambda(y_i^h + h k_3) = \left(\lambda + h \lambda^2 + \frac{1}{2} h^2 \lambda^3 + \frac{1}{4} h^3 \lambda^4\right) y_i^h,$$

and finally

$$y_{i+1}^h = \left(1 + h\lambda + \frac{1}{2} h^2 \lambda^2 + \frac{1}{6} h^3 \lambda^3 + \frac{1}{24} h^4 \lambda^4\right) y_i^h. \tag{7.71}$$

With

$$\Psi(h\lambda) \equiv 1 + h\lambda + \frac{1}{2} h^2 \lambda^2 + \frac{1}{6} h^3 \lambda^3 + \frac{1}{24} h^4 \lambda^4 \tag{7.72}$$

Eq. (7.71) can be written in the form

$$y_{i+1}^h = \Psi(h\lambda) \, y_i^h. \tag{7.73}$$

The function $\Psi(h\lambda)$ is called *stability function*.

Instead of (7.73), the exact solution of the IVP (7.70) satisfies

$$y(x_{i+1}) = e^{\lambda(x_i+h)} = e^{h\lambda} e^{\lambda x_i} = \left(e^{h\lambda}\right) y(x_i). \tag{7.74}$$

Comparing (7.73) and (7.74), we see that $\Psi(h\lambda)$ is the Taylor series for $e^{h\lambda}$ up to and including the term of 4th order. Therefore, if the absolute value of $h\lambda$ is small, the stability function is a good approximation for $e^{h\lambda}$.

If $\lambda > 0$ (i.e., $z \equiv h\lambda > 0$), then $\Psi(z) > 1$. In that case the numerically determined grid function $\{y_i^h\}_{i=0}^{\infty}$ increases, and its growth behavior corresponds with that of the exact solution. It is said: *growing solutions are also integrated growingly.* Since the qualitative behavior of both grid functions is identical, without any restriction on the stability function $\Psi(z)$, the case $\lambda > 0$ is certainly not the interesting one. Moreover, for large positive values of λ, the IVP (7.70) is a mathematically ill-conditioned problem. Even very small perturbations of the initial value can have the effect that the difference between the solutions of the unperturbed and the perturbed IVP strongly increases for $x > x_0$.

Conversely if $\lambda < 0$, then the exact solution decreases, and the IVP (7.70) represents a mathematically well-conditioned problem. But more important is the fact that the numerically determined grid function $\{y_i^h\}_{i=0}^{\infty}$ has the same qualitative growth behavior as $\{y(x_i)\}_{i=0}^{\infty}$ if and only if the stability function satisfies $|\Psi(z)| < 1$. If that is the case, then it is said: *decreasing solutions are also integrated decreasingly.* The stability function (7.72) of the classical RKM is a polynomial of degree 4 for which the following relation is valid:

$$\lim_{z \to -\infty} \Psi(z) = +\infty.$$

Thus, the inequality $|\Psi(z)| < 1$ is not satisfied for all values of z. It is a restriction on the step-size h which guarantees that a special method works qualitatively correct.

The above statements suggest the idea to consider only the case $\lambda < 0$ in the test problem (7.70). Moreover, complex values of λ should also be allowed since in the applications often oscillating solutions occur. This leads us to the following modification of the test problem (7.70):

$$y'(x) = \lambda y(x), \quad y(0) = 1, \quad \lambda \in \mathbb{C} \text{ with } \mathcal{R}e(\lambda) < 0. \tag{7.75}$$

Theorem 7.31 states that not only the classical RKM, but all one-step methods lead to an equation of the form (7.73) if they are applied to the test problem.

Theorem 7.31. *Let the parameters $\Gamma \in \mathbb{R}^{m \times m}$, $\beta \in \mathbb{R}^m$, and $\rho \in \mathbb{R}^m$ of an RKM with m stages be given (see Fig. 7.1). Then, the stability function has the form*

$$\Psi(h\lambda) = 1 + \beta^T h\lambda (I - h\lambda\Gamma)^{-1} \mathbb{1}, \quad \mathbb{1} \equiv (1, 1, \ldots, 1)^T. \tag{7.76}$$

Proof. See, e.g., [9].   ∎

If the RKM is explicit, then $\Gamma^m = 0$. Therefore, the inverse in (7.76) can be represented by the Neumann series for small h. In that case Eq. (7.76) takes the form

$$\Psi(h\lambda) = 1 + \beta^T h\lambda \sum_{j=0}^{m-1} (h\lambda \Gamma)^j \mathbb{1}. \tag{7.77}$$

This shows that, as in the case of the classical RKM, the stability function is a polynomial in $h\lambda$ of degree m. Moreover, it can be shown:

- If an RKM of the order $p \leq 4$ with p stages is applied to the test problem (7.75), then the stability function $\Psi(h\lambda)$ is identical with the first $p + 1$ terms of the Taylor series of $e^{h\lambda}$.
- RKMs of order $p > 4$ have $m > p$ stages (see Fig. 7.7). Therefore, the stability function $\Psi(h\lambda)$ is a polynomial of degree m whose first $p + 1$ terms are identical with the Taylor series of $e^{h\lambda}$. The next terms depend on the special method.

Now we are coming to the following definitions.

Definition 7.32. The set

$$S \equiv \{z \in \mathbb{C} : |\Psi(z) \leq 1|\} \tag{7.78}$$

is called *region of absolute stability* of the one-step method. □

In the case $\mathcal{R}e(\lambda) < 0$, the step-size must be adjusted such that $z = h\lambda$ lies in the region of absolute stability S. Then, the approximations y_i^h are bounded for $i \to \infty$, i.e., the numerical method behaves stable.

Definition 7.33. A one-step method whose region of absolute stability S satisfies the relation

$$S \supset \mathbb{C}^- \equiv \{z \in \mathbb{C} : \mathcal{R}e(z) \leq 0\} \tag{7.79}$$

is called *absolute stable* or *A-stable*. In that case the stability function $\Psi(h\lambda)$ is referred to as *A-compatible*. □

In order to ensure that the numerical determined grid function has the same qualitative behavior as the grid function of the exact solution, the step-size h of an absolute stable one-step method can be chosen without any restriction. Here, the step-size depends only on the required accuracy. However, all *explicit* RKMs are not absolute stable as is stated in Theorem 7.34.

Theorem 7.34. *The region of absolute stability of a consistent m-stage ERK is not empty, bounded, and positioned locally left of the origin.*

Proof. See, e.g., [18]. ■

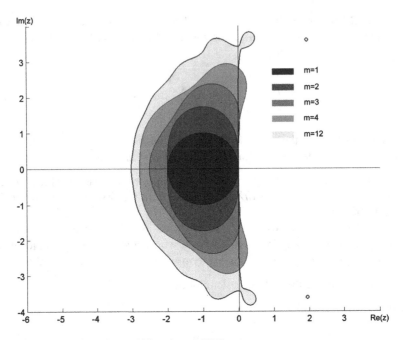

Fig. 7.25 Regions of absolute stability of some ERKs

The region of absolute stability of Euler's (forward) method (see Fig. 7.2) is the circle with the center $z = -1$ and the radius one in the complex plane. Increasing the order of consistency, the region of absolute stability increases as can be seen in Fig. 7.25.

The plotted regions of absolute stability can be determined as follows. Note, all complex numbers whose absolute value is 1 can be represented in the complex plane by $e^{i\theta}, 0 \leq \theta \leq 2\pi$. The stability condition is $|\Psi(z)| \leq 1$, where $\Psi(z)$ is given in (7.77). To compute the boundary of the region of absolute stability, we determine the roots of the equation

$$\Psi(z) - e^{i\theta} = 0 \tag{7.80}$$

for a sequence of θ-values. It is obvious to begin with $\theta = 0$ where $z = 0$. Then, θ is successively increased by a small increment, and in each step the corresponding z is computed with a numerical method for nonlinear algebraic equations (e.g., Newton's method), where as start value the result of the previous step is used. This process is continued until the starting point on the boundary of the region of absolute stability is reached again.

Another strategy for the determination of the region of absolute stability is to lay a fine grid on a large part of the left half-plane. Then, in each grid point the value of the stability function $\Psi(z)$ is computed. Those grid points z_{ij} are marked as belonging to the region of absolute stability which satisfy $|\Psi(z_{ij})| < 1$.

Table 7.3 Stability intervals of some ERKs

Order p	1	2	3	4	5
Stability interval	$[-2,0]$	$[-2,0]$	$[-2.51,0]$	$[-2.78,0]$	$[-3.21,0]$

Fig. 7.26 Midpoint rule

$$
\begin{array}{c|c}
\frac{1}{2} & \frac{1}{2} \\
\hline
& 1
\end{array}
$$

A measure for the magnitude of the region of absolute stability is also the so-called stability interval. For the ERKs considered in Fig. 7.25 the corresponding stability intervals are presented in Table 7.3.

Let us now study the region of absolute stability of *implicit* one-step methods. The application of the trapezoidal method (see Fig. 7.4) to the test problem (7.75) gives

$$
y_{i+1}^h = y_i^h + \frac{h}{2}\left(\lambda y_i^h + \lambda y_{i+1}^h\right), \text{ i.e. } \quad y_{i+1}^h = \frac{1 + \frac{1}{2}h\lambda}{1 - \frac{1}{2}h\lambda}\, y_i^h \equiv \Psi(h\lambda)y_i^h. \quad (7.81)
$$

Obviously, the stability function $\Psi(z)$ is a rational function. It satisfies

$$
|\Psi(z)| = \left|\frac{2+z}{2-z}\right| < 1 \quad \text{for all } z \text{ with } \mathcal{R}e(z) < 0,
$$

since for $\mathcal{R}e(z) < 0$, the absolute value of the real part of the numerator is always smaller than the absolute value of the real part of the denominator, whereas the imaginary parts differ only in sign. Thus, the region of absolute stability contains the whole left half-plane, i.e., the trapezoidal method is A-stable.

Another example of an implicit RKM is the midpoint rule. The associated Butcher diagram is given in Fig. 7.26.

The application of the midpoint rule to the test problem (7.75) yields

$$
k_1 = \lambda\left(y_i^h + \frac{1}{2}hk_1\right), \text{ i.e. } \quad k_1 = \frac{\lambda}{1 - \frac{1}{2}h\lambda}\, y_i^h.
$$

It follows that

$$
y_{i+1}^h = y_i^h + h\,k_1 = \frac{1 + \frac{1}{2}h\lambda}{1 - \frac{1}{2}h\lambda}\, y_i^h = \Psi(h\lambda)y_i^h.
$$

The midpoint rule is again A-stable since it has the same stability function as the trapezoidal method.

Now, the question arises whether all implicit RKMs are A-stable. In answering this question, the concept of the Padé approximation of a given function plays an important role.

Definition 7.35. Let $f(z)$ be an analytical function in a neighborhood of the point $z = 0$. Then, the rational function

$$R_{jk}(z) = \frac{P_{jk}(z)}{Q_{jk}(z)} = \frac{\displaystyle\sum_{l=0}^{k} a_l z^l}{\displaystyle\sum_{l=0}^{j} b_l z^l}, \quad b_0 = 1,$$

is called a Padé approximation of $f(z)$ with index (j, k), if

$$R_{jk}^{(l)}(0) = f^{(l)}(0), \quad l = 0, \ldots, j + k. \tag{7.82}$$

□

For the implicit RKMs studied in Sect. 7.6, the following can be stated.

Theorem 7.36. *The stability function $\Psi(z)$ of the following implicit m-stage RKMs is a Padé approximation of e^z, namely, for:*

- *The Gauss method with the index (m, m)*
- *The Radau IA method with the index $(m, m-1)$*
- *The Radau IIA method with the index $(m, m-1)$*
- *The Lobatto IIIA method with the index $(m-1, m-1)$*
- *The Lobatto IIIB method with the index $(m-1, m-1)$*
- *The Lobatto IIIC method with the index $(m, m-2)$*

Proof. See, e.g., [26].   ■

Therefore, the proof of the A-stability of an implicit RKM requires the verification of the A-compatibility (see Definition 7.33) of the corresponding Padé approximation. This connection is established in Theorem 7.37.

Theorem 7.37. *All Padé approximations with index (j, k), for which the relation $j - 2 \le k \le j$ holds, are A-compatible.*

Proof. See, e.g., [26].   ■

Consequently, the implicit RKMs mentioned in Theorem 7.36 are A-stable.

In practice, one can very often find IVPs which cannot be solved by ERKs. The reason is the occurrence of different time scales in the mathematical model. Such IVPs are called *stiff*. From a mathematical point of view, there are various definitions of the concept of stiffness which describe the problem on a high theoretical level more or less correct (see, e.g., [19]). A more pragmatic definition of stiffness is related to the behavior of Euler's (forward) method which is representative for all *explicit* RKMs (see also [2]).

Definition 7.38. The IVP (7.5) is referred to as stiff in the interval $[x_0, x_N]$ if the step-size needed to maintain absolute stability of Euler's (forward) method is much smaller than the step-size required to compute the solution accurately.   □

Thus, if an ERK with an automatic step-size control (see Sect. 7.8) is used and the step-size tends to zero (i.e., the method cannot go forward), then the problem is generally stiff. In such a case the A-stable IRKs must be applied whose region of absolute stability is the entire left half-plane. However, one should always try to solve a given IVP by an ERK because these methods do not need the costly solution of systems of nonlinear algebraic equations.

7.10 Exercises

Exercise 7.1. Transform the following systems of higher-order ODEs into equivalent systems of first-order ODEs:

(1) $v^{(4)}(x) - a(x)u'(x) = f(x)$, $u''(x) + b(x)v(x) = g(x)$.
(2) $v^{(4)}(x) - a(x)u''(x) = f(x)$, $u''(x) + b(x)v(x) = g(x)$.

Exercise 7.2. Formulate the Butcher diagram for the following RKM:

$$y_{i+2/3} = y_i + \frac{h}{3}\left[f(x_i, y_i) + f\left(x_i + \frac{2}{3}h, y_i + \frac{2}{3}hf(x_i, y_i)\right)\right],$$

$$y_{i+1} = y_i + \frac{h}{4}\left[f(x_i, y_i) + 3f\left(x_i + \frac{2}{3}h, y_{i+2/3}\right)\right].$$

Show that this RKM has at least the order of consistency $p = 3$.

Exercise 7.3. Let the following IVP be given which has to be solved by Euler's (backward) method:

$$\varphi''(x) = \sin(\varphi(x)), \quad \varphi(x_0) = \varphi_0, \ \varphi'(x_0) = \varphi_0', \quad x_0 \le x \le T.$$

1. Transform this IVP into a system of two first-order equations. Which system of nonlinear algebraic equations must be solved in each step of Euler's (backward) method?
2. Write a MATLAB code for this task, where the interval endpoints x_0, T, the initial values φ_0, φ_0', and the step-size h have to be prescribed. Use the MATLAB-function fsolve to solve the associated system of nonlinear algebraic equations.

Exercise 7.4. Consider the IVP for a linear inhomogeneous ODE:

$$y'(x) = f(x, y(x)), \quad f(x, y) = py + q, \quad y(x_0) = a, \quad p, q \in \mathbb{R}, \ p \neq 0.$$

Assume that Euler's (forward) method

$$y_{i+1} = y_i + h f(x_i, y_i), \quad i = 0, 1, \ldots, \tag{7.83}$$

is used to solve this IVP, where $y_0 = a$ and $x_i = x_0 + i h$, $i = 0, 1, \ldots$.

Fig. 7.27 RKMs of order 3

ρ_1				
ρ_2	1			
ρ_3	$\dfrac{1}{2}$	$\dfrac{1}{2}$		
ρ_4	$\dfrac{1}{2}$	$\dfrac{1}{2}$	α	
	β_1	β_2	β_3	β_4

1. Determine the exact solution of the IVP.
2. Show that

$$y_i = c\,(1 + ph)^i - \frac{q}{p}, \quad i = 1, 2, \ldots, \tag{7.84}$$

is the solution of the difference equation (7.83). Here, $c \in \mathbb{R}$ is an arbitrary constant.
3. Determine the constant c such that (7.84) satisfies the initial condition $y_0 = a$, too.

Exercise 7.5. Consider the autonomous IVP:

$$y'(x) = f(y(x)), \quad y(x_0) = y_0, \quad x_0 \le x \le T.$$

Assume that the function f is sufficiently smooth. Use the order conditions given in Theorem 7.19 to construct RKMs of the order $p = 3$ which are given by the Butcher diagram in Fig. 7.27.

1. Show that for each α which satisfies $\alpha(\alpha + 1) \neq 0$, there exist a uniquely determined set of parameters β_j, $j = 1, \ldots, 4$, such that the RKM given by Fig. 7.27 has the order $p = 3$.
2. What is the appropriate choice of α?

Exercise 7.6. Let the following IVP be given:

$$y'(x) = f(x, y(x)), \quad y(x_0) = y_0, \quad x_0 \le x \le T, \tag{7.85}$$

where

$$f(x, y) = -10(y - \arctan(x)) + \frac{1}{1 + x^2}, \quad x_0 = -4, \quad y_0 = 2, \quad T = 6.$$

Fig. 7.28 A 2-stage DIRK

$$
\begin{array}{c|cc}
0 & & \\
1 & \dfrac{1}{2} & \dfrac{1}{2} \\
\hline
& \dfrac{1}{2} & \dfrac{1}{2}
\end{array}
$$

1. Determine the exact solution $y(x)$ of the IVP (7.85).
2. Transform (7.85) into an autonomous IVP

$$
z'(x) = g(z(x)), \quad z(x_0) = z_0. \tag{7.86}
$$

3. Integrate (7.86) with Euler's (forward) method and the different step-sizes $h_n = (T - x_0)/n$, $n = 48, 50, 55, 128$. Use these results $z_i^{h_n}$ and the relation between (7.85) and (7.86) to compute approximations $y_i^{h_n}$ for $y(x_i)$. Represent graphically $y_i^{h_n}$ and $y(x_i)$, $i = 1, 2, \ldots, n$.
4. For each of the abovementioned step-sizes h_n, explain theoretically the behavior of the computed approximations $y_i^{h_n}$.
5. Determine theoretically the greatest possible step-size $h > 0$ such that the associated approximation computed by Euler's (forward) method is bounded on $[x_0, \infty)$.

Exercise 7.7. An interesting fact of RKMs is that the local discretization error depends on the type of the ODE as well as on the solution of the IVP. To demonstrate this, show:

1. The function $y(x) = (x + 1)^2$ solves the following two IVPs

$$
y'(x) = 2(x + 1), \quad y(0) = 1, \quad \text{and} \quad y'(x) = \frac{2y(x)}{x + 1}, \quad y(0) = 1.
$$

2. The first IVP is integrated exactly by Heun's method.
3. The second IVP is not integrated exactly by Heun's method.

Exercise 7.8. The Butcher diagram given in Fig. 7.28 defines a diagonally implicit RKM (DIRK).

1. Write down the difference equation of this RKM.
2. Determine the order of consistency of this method.

Exercise 7.9. In Fig. 7.29 the Butcher diagram of an ERK of the order of consistency $p = 3$ is given.

1. Construct an ERK of order $p = 2$ which does not require additional function values and is embedded into the above Butcher diagram presented in Fig. 7.29.
2. Use both ERKs to develop a MATLAB implementation where the step-size is controlled by the embedding principle (see Sect. 7.8).

Fig. 7.29 ERK of order 3

$$
\begin{array}{c|ccc}
0 & & & \\
\dfrac{1}{2} & \dfrac{1}{2} & & \\
1 & -1 & 2 & \\
\hline
& \dfrac{1}{6} & \dfrac{2}{3} & \dfrac{1}{6}
\end{array}
$$

Fig. 7.30 Nørsett's method;
$\alpha = \pm 1/(2\sqrt{3})$

$$
\begin{array}{c|cc}
\dfrac{1}{2} + \alpha & \dfrac{1}{2} + \alpha & \\
\dfrac{1}{2} - \alpha & -2\alpha & \dfrac{1}{2} + \alpha \\
\hline
& \dfrac{1}{2} & \dfrac{1}{2}
\end{array}
$$

Exercise 7.10. Determine the stability interval of the 2-stage Nørsett method given in Fig. 7.30.

Exercise 7.11. Let the following IVP be given:

$$y'(x) = \lambda(y(x) - x) + 1, \quad y(0) = 1, \quad 0 \le x \le 10.$$

1. Determine the exact solution of the IVP.
2. Write a MATLAB program which is based on the MATLAB code ode45, and solve the above IVP on the interval $[0, 10]$ for

$$\lambda = -10, -20, -30, -40, -50, -100.$$

   Use the following parameters:
```
                      options =
   odeset('RelTol',1e-8,'AbsTol',1e-8,'stats','on','refine',1);
```
3. Do the same using the MATLAB code ode15s instead of ode45.
4. Give the number of function calls needed by each code and the corresponding errors of the numerical results. Explain the different behaviors of the two codes.

Exercise 7.12. Given the IVP

$$y'(x) = \sqrt[5]{y(x)}, \quad y(0) = 0$$

and the associated exact solution

$$y(x) = \left(\frac{4x}{5}\right)^{\frac{5}{4}},$$

explain that Euler's (forward) method is not appropriate to compute an approximation of $y(x)$.

Exercise 7.13. In calm water a tugboat goes linearly with constant speed to North-East and has taken into tow a small boat. The tow rope has the length L and is always in constant tension. Compute the course of the small boat.

If we assume that the tangent to the trajectory of the course is represented by the tow rope, the problem leads to the IVP

$$y'(x) = \frac{t - y(x)}{t - x}, \quad y(x_0) = y_0. \tag{7.87}$$

Here, (t, t) is the position of the tugboat and (x, y) is the position of the small boat. Both are linked by the relation

$$(t - y)^2 + (t - x)^2 = L^2. \tag{7.88}$$

For our experiments, let us choose the following data:

$$L = 10, \quad x_l = x_0 = -\frac{L}{\sqrt{2}}, \quad y_0 = \frac{L}{\sqrt{2}}, \quad x_l \le x \le x_r = 16.$$

1. The exact solution of (7.87) is

$$s + \int \frac{\sqrt{2L^2 - s^2}}{s} \, ds = -2x, \quad s \equiv y - x, \tag{7.89}$$

where the constant of integration has to be determined by the initial condition. Derive the IVP (7.87) and compute the exact solution (7.89).
2. Solve the IVP (7.87) numerically by Euler's (forward) method, Heun's method, and the classical Runge-Kutta method. For all methods, let the step-size h be constant, more precisely $h = 0.25$. Display the numerically determined courses of the tugboat and the small boat graphically. The result is presented in Fig. 7.31. *Hint:* Solve Eq. (7.88) for t and substitute this expression into (7.87).
3. Solve the problem as outlined in 2. If the tugboat goes on a circle with the center $(0, 0)$ and the radius R, solve this problem as outlined in 2. Define yourself the initial position of the small boat.

Exercise 7.14. Given the following IVP

$$y'(x) = \frac{3x^2}{3y(x)^2 - 4}, \quad y(0) = 0.$$

1. Show that the solution of the above IVP satisfies $y^3 - 4y = t^3$. Explain why the solution $y(x)$ cannot be continued outside the interval $|t| < T$, where $T \approx 1.454834120$.

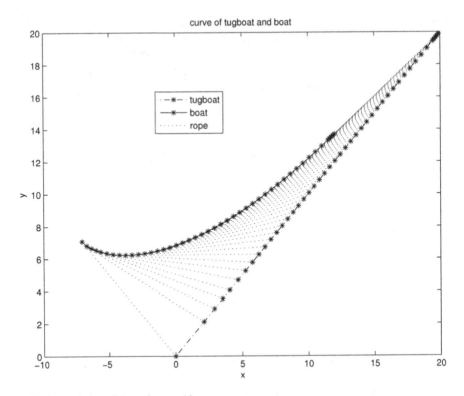

Fig. 7.31 Solution of the tugboat problem

2. Determine approximately by numerical experiments the greatest possible interval I on which the IVP

$$\left(y(x)^2 - 1\right) y'(x) = t^2, \quad y(0) = 0,$$

has a unique solution. Try to obtain I analytically.

Exercise 7.15. Given the IVP

$$y'(x) = \sqrt{\frac{y(x)^2}{1 + y(x)^2}}, \quad y(x_0) = y_0.$$

1. For each of the following initial conditions

     (i) $y(1) = 1$,   (ii) $y(1) = -1$, and   (iii) $y(1) = 0$,

determine the exact solution $y(x)$. Use this solution to compute the numerical value of $y(3)$. Analyze the behavior of $y(x)$ for $x \to \pm\infty$ and determine the corresponding asymptotes of the solution. Display the solutions of the IVPs graphically on the interval $[-2, 4]$.

2. Use Euler's (forward) method to approximate the exact solution of the above IVPs and proceed as described in 1.
3. Discuss the advantages and drawbacks of the approaches considered in 1. and 2.

Exercise 7.16. The step-size control on the basis of the Runge principle (see Sect. 7.8) is to be tested and examined. Use Newton's 3/8 rule as basic method.

1. Write a MATLAB code for Newton's 3/8 rule with step-size control.
2. Use the van der Pol equation

$$y''(x) - \left(1 - y(x)^2\right) y'(x) + y(x) = 0$$

as test problem.

Compare your code with the MATLAB function ode45. How many integration steps and function calls are required to solve the test problem? How many rejected step-sizes occur?

Exercise 7.17. The quadrature formula

$$Q_f[-1, 1] \equiv \frac{3}{2} f\left(-\frac{1}{3}\right) + \frac{1}{2} f(1) \tag{7.90}$$

can be used to approximate the integral

$$\int_{-1}^{1} f(s) ds. \tag{7.91}$$

1. Show that (7.90) is exact for all polynomials of degree less than or equal to 2.
2. Derive from (7.90) a quadrature formula $Q_f[x, x + h]$ for the integral

$$\int_{x}^{x+h} f(s) ds. \tag{7.92}$$

3. Use (7.92) to develop a Runge-Kutta method for the numerical solution of the scalar IVP:

$$y'(x) = f(x, y(x)), \quad y(x_0) = y_0.$$

To do this (see Sect. 7.3), apply the quadrature formula $Q_f[x, x + h]$ on

$$y(x + h) = y(x) + \int_{x}^{x+h} f(\tau, y(\tau)) d\tau.$$

In this way the IRK presented in Fig. 7.32 results. Develop this IRK and determine its order of consistency.

Fig. 7.32 A 2-stage IRK

$$
\begin{array}{c|cc}
\dfrac{3}{4} & \dfrac{5}{12} & -\dfrac{1}{12} \\[2ex]
1 & \dfrac{3}{4} & \dfrac{1}{4} \\[2ex]
\hline
 & \dfrac{3}{4} & \dfrac{1}{4}
\end{array}
$$

Fig. 7.33 A 2-stage ERK

$$
\begin{array}{c|cc}
0 & & \\[1ex]
\rho & \gamma & \\[1ex]
\hline
 & \beta_1 & \beta_2
\end{array}
$$

Exercise 7.18. The equation

$$
\frac{x^2}{a^2} + \frac{y^2}{b^2} = 1, \quad a, b > 0 \tag{7.93}
$$

defines an ellipse. Make a qualitatively correct drawing of this ellipse.

1. Develop a first-order ODE such that the solution of the corresponding IVP

$$
y'(x) = f(x, y(x)), \quad y(0) = b, \tag{7.94}
$$

satisfies Eq. (7.93). In addition, develop a system of two *linear* first-order ODEs such that the solution of the corresponding IVP

$$
y'(t) = A\, y(t), \quad y(0) = (a, 0)^T, \tag{7.95}
$$

satisfies Eq. (7.93), too. Here, $A \in \mathbb{R}^{2 \times 2}$ and $y(t) \equiv (x(t), y(t))^T$.
2. Solve the IVPs (7.94) and (7.95) by Euler's (forward) method. Which problem is more suitable from a numerical point of view to compute the solution curve which is determined by (7.93)?

Exercise 7.19. Consider the IVP

$$
y'(x) = a\left(y(x) + y(x)^3\right), \quad x \geq 0, \quad y(0) = 1, \quad a > 0. \tag{7.96}
$$

1. Formulate conditions on a which guarantee that the corresponding solution $y(x; a)$ of (7.96) is defined on $[0, 1]$.
2. By the use of experiments, investigate the behavior of Euler's (forward) method when it is applied to (7.96) on an interval $[0, \tau]$, $\tau > 1$.

Exercise 7.20. Let the Butcher diagram in Fig. 7.33 be given.

1. Determine the coefficients β_1, β_2 and γ such that the local discretization error can be represented in the form

$$\delta(x, y(x); h) = C(\rho, x) h^2 + O(h^3),$$

where $C : (0, 1] \times [x_0, T] \to \mathbb{R}$ is a suitable function.

2. Assume that there exist constants $M_k > 0$, $k = 0, \ldots, 3$, and $M_y > 0$ such that

$$\|y^{(k)}(x)\| \le M_k, \quad k = 0, \ldots, 3, \quad \text{and} \quad \|f_y(x, y(x))\| \le M_y,$$

for all $x \in [x_0, T]$. Show that for all $\rho \in (0, 1]$ and all $x \in [x_0, T]$, it holds

$$|C(\rho, x)| \le K(\rho) \equiv \frac{1}{2} \left[M_3 + M_y M_2 \right] \left| \frac{1}{3} - \frac{1}{2} \rho \right| + \frac{1}{6} M_y M_2. \tag{7.97}$$

3. Determine an optimum value $\rho \in (0, 1]$ by which the upper limit in (7.97) is minimized and write down the associated Butcher diagram and the ESM.

References

1. Abramowitz, M., Stegun, I.A.: Handbook of Mathematical Functions. Dover, New York (1972)
2. Ascher, U.M., Petzold, L.R.: Computer Methods for Ordinary Differential Equations and Differential-Algebraic Equations. SIAM, Philadelphia (1998)
3. Butcher, J.C.: Coefficients for the study of Runge-Kutta integration processes. J. Aust. Math. Soc. 3, 185–201 (1963)
4. Butcher, J.C.: On Runge-Kutta processes of high order. J. Aust. Math. Soc. 4, 179–194 (1964)
5. Butcher, J.C.: On the attainable order of Runge-Kutta methods. Math. Comput. 19, 408–417 (1965)
6. Butcher, J.C.: Numerical Methods for Ordinary Differential Equations. Wiley, Chichester (2003)
7. Chipman, F.H.: A-stable Runge-Kutta processes. BIT 11, 384–388 (1971)
8. Dahlquist, G.: A special stability problem for linear multistep methods. BIT 3, 27–43 (1963)
9. Dekker, K., Verwer, J.G.: Stability of Runge-Kutta Methods for Stiff Nonlinear Differential Equations. North-Holland, Amsterdam (1984)
10. Dormand, J.R., Prince, P.J.: A family of embedded Runge-Kutta formulae. J. Comput. Appl. Math. 6, 19–26 (1980)
11. Ehle, B.L.: High order A-stable methods for the numerical solution of systems of DEs. BIT 8, 276–278 (1968)
12. Euler, L.: Institutionum calculi integralis. Volumen Primum, Opera Omnia XI (1768)
13. Fehlberg, E.: Classical fifth-, sixth-, seventh-, and eighth order Runge-Kutta formulas with step size control. Technical report 287, NASA (1968)
14. Fehlberg, E.: Low-order classical Runge-Kutta formulas with step size control and their application to some heat transfer problems. Technical report 315, NASA (1969)
15. Hairer, E., Wanner, G.: On the Butcher group and general multi-value methods. Computing 13, 1–15 (1974)
16. Hairer, E., Nørsett, S.P., Wanner, G.: Solving Ordinary Differential Equations, vol. I. Springer, Berlin (1993)
17. Hartman, P.: Ordinary Differential Equations. Birkhäuser, Boston/Basel/Stuttgart (1982)

18. Hermann, M.: The numerical treatment of perturbed bifurcation in boundary value problems. In: Wakulicz, A. (ed.) Computational Mathematics, pp. 391–405. PWN-Polish Scientific Publishers, Warszawa (1984)
19. Hermann, M.: Numerik gewöhnlicher Differentialgleichungen. Anfangs- und Randwertprobleme. Oldenbourg Verlag, München und Wien (2004)
20. Hermann, M.: Numerische Mathematik, 3rd edn. Oldenbourg Verlag, München (2011)
21. Heun, K.: Neue Methode zur approximativen Integration der Differentialgleichungen einer unabhängigen Veränderlichen. Zeitschr. für Math. u. Phys. **45**, 23–38 (1900)
22. Kaiser, D.: Erzeugung von eingebetteten Runge-Kutta-Verfahren hoher Ordnung, z.b. 10(8), 12(10) und 14(12) durch Modifikation bekannter Verfahren ohne Erhöhung der Stufenanzahl. JENAER SCHRIFTEN ZUR MATHEMATIK UND INFORMATIK 03, Friedrich Schiller University at Jena (2013)
23. Kutta, W.: Beitrag zur näherungsweisen Integration totaler Differentialgleichungen. Zeitschr. für Math. u. Phys. **46**, 435–453 (1901)
24. Runge, C.: Über die numerische Auflösung von Differentialgleichungen. Math. Ann. **46**, 167–178 (1895)
25. Strehmel, K., Weiner, R.: Numerik gewöhnlicher Differentialgleichungen. Teubner Verlag, Stuttgart (1995)
26. Strehmel, K., Weiner, R., Podhaisky, H.: Numerik gewöhnlicher Differentialgleichungen. Springer Spektrum, Wiesbaden (2012)
27. Verner, J.H.: Explicit Runge-Kutta methods with estimates of the local truncation error. SIAM J. Numer. Anal. **15**, 772–790 (1978)

Chapter 8
Shooting Methods for Linear Boundary Value Problems

8.1 Two-Point Boundary Value Problems

In the previous chapter, we have presented numerical methods for the approximate solution of IVPs of the form (7.5) consisting of n *nonlinear* first-order ODEs and a set of n initial conditions. In this chapter, we will replace the initial conditions by n *linear* two-point boundary conditions. Moreover, let us assume that the ODEs are *linear*, too. The resulting problem is

$$\mathscr{L}y(x) \equiv y'(x) - A(x)y(x) = r(x), \quad a \leq x \leq b, \tag{8.1a}$$

$$\mathscr{B}y(x) \equiv B_a y(a) + B_b y(b) = \beta, \tag{8.1b}$$

where $A(x), B_a, B_b \in \mathbb{R}^{n \times n}$ and $r(x), \beta \in \mathbb{R}^n$. We assume that $A(x)$ and $r(x)$ are sufficiently smooth.

Problems of the form (8.1a) and (8.1b) are called (linear) *two-point boundary value problems* (BVPs). If components of the solution $y(x)$ are prescribed at more than two points, we have a multipoint boundary value problem. IVPs are typical for evolution problems, and x represents the time. In contrast, boundary value problems are used to model static problems (e.g., elasto-mechanical problems) since the solution depends on the state of the system in the past (at the point a) as well as on the state of the system in the future (at the point b). In general, the variable x in a BVP does not describe the time, i.e., no dynamics is inherent in the given problem.

A special case of linear boundary value conditions are *partially separated* boundary value conditions. Here, p $(<n)$ equations of the linear n-dimensional algebraic system $\mathscr{B}y = \beta$ are only defined at one endpoint. Without loss of generality, let us assume that this endpoint is $x = a$. Then, after a possible rearrangement of the equations, the matrices B_a and B_b can be written in the form

M. Hermann and M. Saravi, *A First Course in Ordinary Differential Equations: Analytical and Numerical Methods*, DOI 10.1007/978-81-322-1835-7_8, © Springer India 2014

$$B_a = \begin{pmatrix} B_a^{(1)} \\ B_a^{(2)} \end{pmatrix}, \quad B_b = \begin{pmatrix} 0 \\ B_b^{(2)} \end{pmatrix}, \tag{8.2}$$

where $B_a^{(1)} \in \mathbb{R}^{p \times n}$ and $B_a^{(2)}, B_b^{(2)} \in \mathbb{R}^{q \times n}$, $p + q = n$.

We have another special case when in each of the n linear boundary equations, the components of the solution occur at only one of the two endpoints a and b. Then, after a possible rearrangement of the equations, the boundary matrices are

$$B_a = \begin{pmatrix} B_a^{(1)} \\ 0 \end{pmatrix}, \quad B_b = \begin{pmatrix} 0 \\ B_b^{(2)} \end{pmatrix}. \tag{8.3}$$

Boundary conditions with boundary matrices of the form (8.3) are called *completely separated* boundary conditions.

If none of the abovementioned special cases applies, the boundary conditions are denoted as *non-separated*.

A necessary condition for the solvability of the BVP (8.1a) and (8.1b) is that the rank of the matrix $(B_a | B_b) \in \mathbb{R}^{n \times 2n}$ is n. Therefore, in this chapter we assume that the relation

$$\text{rank}(B_a | B_b) = n \tag{8.4}$$

is satisfied.

Example 8.1. Given the linear BVP

$$\dot{y}_1(x) = 2y_1(x) + y_2(x), \quad y_1(1) = e,$$

$$\dot{y}_2(x) = y_1(x) + 2y_2(x), \quad y_2(0) = -1.$$

The exact solution is $y_1(x) = e^x$ and $y_2(x) = -e^x$. Since the ODEs are homogeneous, we have $r(x) \equiv 0$ in (8.1a) and (8.1b). Thus, in matrix notation the ODEs are

$$\dot{y}(x) - \begin{pmatrix} 2 & 1 \\ 1 & 2 \end{pmatrix} y(x) = 0,$$

with $y(x) \equiv (y_1(x), y_2(x))^T$. Using the boundary matrices

$$B_a = \begin{pmatrix} 0 & 1 \\ 0 & 0 \end{pmatrix} \quad \text{and} \quad B_b = \begin{pmatrix} 0 & 0 \\ 1 & 0 \end{pmatrix},$$

the separated boundary conditions can be written in the form

$$\begin{pmatrix} 0 & 1 \\ 0 & 0 \end{pmatrix} y(0) + \begin{pmatrix} 0 & 0 \\ 1 & 0 \end{pmatrix} y(1) = \begin{pmatrix} -1 \\ e \end{pmatrix}.$$

Setting $p = q \equiv 1$, $B_a^{(1)} \equiv (0,1)$, $B_b^{(2)} \equiv (1,0)$, and $B_a^{(2)} = B_b^{(1)} \equiv (0,0)$, we see that the boundary matrices have the structure (8.3).

The assumption (8.4) is satisfied since

$$(B_a|B_b) = \left(\begin{array}{cc|cc} 0 & 1 & 0 & 0 \\ 0 & 0 & 1 & 0 \end{array} \right) \in \mathbb{R}^{2\times 4}.$$

$\square$

Let $Y(x;a) \in \mathbb{R}^{n\times n}$ be a fundamental matrix of the homogeneous part of the ODE in (8.1a) and (8.1b), i.e., $Y(x;a)$ satisfies

$$\mathscr{L}Y(x;a) = 0, \quad a \le x \le b, \quad Y(a;a) = Y^a \in \mathbb{R}^{n\times n}, \quad \det(Y^a) \ne 0. \quad (8.5)$$

Moreover, let $v(x;a) \in \mathbb{R}^n$ be a particular solution of the inhomogeneous ODE in (8.1a) and (8.1b) which satisfies

$$\mathscr{L}v(x;a) = r(x), \quad a \le x \le b, \quad v(a;a) = v^a \in \mathbb{R}^n. \quad (8.6)$$

The second argument in $Y(x;a)$ and $v(x;a)$ is used to point out that these functions are defined by initial values at $x = a$. Since the ODE in (8.1a) and (8.1b) is linear, the principle of superposition applies. Therefore, the general solution of the ODE can be written in the form

$$y(x) = Y(x;a)c + v(x;a), \quad (8.7)$$

where the vector $c \in \mathbb{R}^n$ is not determined yet. To determine c we have to consider the boundary conditions in (8.1a) and (8.1b). Substituting the ansatz (8.7) into the boundary conditions, we obtain the following n-dimensional system of linear algebraic equations

$$(B_a Y(a;a) + B_b Y(b;a)) c = \beta - B_a v(a;a) - B_b v(b;a). \quad (8.8)$$

Now, in Theorem 8.2 the condition is formulated which guarantees a unique solution of the BVP (8.1a) and (8.1b).

Theorem 8.2. *Suppose that $A \in C^s([a,b], \mathbb{R}^{n\times n})$, $r \in C^s([a,b], \mathbb{R}^{n\times n})$, and $Y(x;a)$ is a fundamental matrix. Then, the BVP* (8.1a) *and* (8.1b) *possesses a unique solution $y \in C^{s+1}([a,b], \mathbb{R}^n)$ if and only if the matrix*

$$M \equiv \mathscr{B}Y(x;a) = B_a Y(a;a) + B_b Y(b;a) \quad (8.9)$$

is nonsingular.

Proof. The claim follows immediately from (8.8). ∎

We will demonstrate the statement of Theorem 8.2 by an example.

Example 8.3. Given the following BVP for a scalar second-order ODE on the interval $[0, b]$

$$z''(x) = -z(x), \quad z(0) = 0, \quad z(b) = z_b. \tag{8.10}$$

We set

$$y = (y_1, y_2)^T \equiv (z, z')^T, \quad y_1' = y_2, \quad y_2' = -y_1$$

and formulate the problem in matrix notation

$$y'(x) - \begin{pmatrix} 0 & 1 \\ -1 & 0 \end{pmatrix} y(x) = 0,$$

$$\begin{pmatrix} 1 & 0 \\ 0 & 0 \end{pmatrix} \begin{pmatrix} y_1(0) \\ y_2(0) \end{pmatrix} + \begin{pmatrix} 0 & 0 \\ 1 & 0 \end{pmatrix} \begin{pmatrix} y_1(b) \\ y_2(b) \end{pmatrix} = \begin{pmatrix} 0 \\ z_b \end{pmatrix}.$$

Since the ODE is linear and homogeneous, and has constant coefficients, the fundamental matrix $Y(x; a)$ can be determined with the usual ansatz on the basis of the exponential function. If we use the initial condition $Y(0; 0) = I$, it follows

$$Y(x; 0) = \begin{pmatrix} \cos(x) & \sin(x) \\ -\sin(x) & \cos(x) \end{pmatrix}.$$

Now, the Matrix M is given by (8.9)

$$M = \begin{pmatrix} 1 & 0 \\ 0 & 0 \end{pmatrix} \begin{pmatrix} 1 & 0 \\ 0 & 1 \end{pmatrix} + \begin{pmatrix} 0 & 0 \\ 1 & 0 \end{pmatrix} \begin{pmatrix} \cos(x) & \sin(x) \\ -\sin(x) & \cos(x) \end{pmatrix} = \begin{pmatrix} 0 & 0 \\ \cos(b) & \sin(b) \end{pmatrix}.$$

Thus, there is only a (unique) solution of the linear BVP (8.10) if the right endpoint of the interval $[0, b]$ satisfies $b \neq k\pi, k \in \mathbb{N}$. □

The above example shows very impressively that even for linear BVPs (in contrast to IVPs), the smoothness of the right-hand side of the ODE does not guarantee the existence of a solution.

8.2 Simple Shooting Method

The basic principle of the so-called *shooting methods* can be described as follows. The solution of the BVP (8.1a) and (8.1b) is constructed by the numerical integration of an associated IVP

$$\mathcal{L}y(x) = r(x), \quad y(a) = y_a, \quad x \in [a, b]. \tag{8.11}$$

The initial vector y_a is still unknown and has to be determined such that the trajectory of the IVP (8.11) fulfills the given boundary conditions, too.

In the *nonlinear case*, the determination of y_a is realized by successive corrections of an estimated initial vector $\tilde{y}_a$ in analogy to the systematic trial and error procedure of the artillery. From this, the name of the method results: *shooting method*.

In the *linear case* y_a can be calculated from the linear algebraic system (8.8), without trial and error, in one step. Let us assume that the following $n + 1$ IVPs are uniquely solvable and there exists good numerical software to treat these IVPs (see Chap. 7):

- One IVP for the particular solution $v(x; a)$

$$\mathcal{L}v(x; a) = r(x), \quad v(a; a) = v^a \in \mathbb{R}^n \text{ (arbitrary)}, \tag{8.12}$$

- n IVPs for the columns $y_i(x; a)$ of the fundamental matrix
  $Y(x; a) = \left(y_1(x; a) | y_2(x; a) | \cdots | y_n(x; a) \right)$

$$\mathcal{L}y_i(x; a) = 0, \quad y_i(a; a) = y_i^a \in \mathbb{R}^n, \quad i = 1, \ldots, n. \tag{8.13}$$

These IVPs can be written in a compact form as

$$\mathcal{L}Y(x; a) = 0, \quad Y(a; a) = Y^a \in \mathbb{R}^{n \times n} \text{ (nonsingular)}.$$

Following (8.7) and (8.8), we get the n-dimensional linear algebraic system which determines the vector $c \in \mathbb{R}^n$

$$Mc = q, \tag{8.14}$$

where $M \equiv B_a Y(a; a) + B_b Y(b; a)$ and $q \equiv \beta - B_a v(a; a) - B_b v(b; a)$.

If M is nonsingular and well conditioned, the system (8.14) can be solved by numerical methods for linear algebraic systems (e.g., Gaussian elimination). Then, using formula (8.7) the missing initial vector y_a is computed as

$$y_a = Y^a c + v^a, \tag{8.15}$$

and, finally, the associated IVP (8.11) can be integrated numerically to approximate the solution $y(x)$ of the BVP (8.1a) and (8.1b) at various points of interest $x_k \in [a, b]$.

The numerical method described above is the simplest shooting technique. It is called *simple shooting method*.

Example 8.4 (See [9], pages 91–111). Given the following scalar linear BVP

$$y''(x) = -\frac{3\lambda}{(\lambda + x^2)^2}\, y(x), \quad \lambda \in \mathbb{R}, \quad x \in [-1, 1],$$

$$(8.16)$$

$$y(1) = -y(-1) = \frac{1}{\sqrt{\lambda + 1}}.$$

At first we write (8.16) as a system of two first-order ODEs:

$$y_1'(x) = y_2(x), \quad y_2'(x) = -\frac{3\lambda}{(\lambda + x^2)^2}\, y_1(x),$$

$$\begin{pmatrix} 1 & 0 \\ 0 & 0 \end{pmatrix} \begin{pmatrix} y_1(-1) \\ y_2(-1) \end{pmatrix} + \begin{pmatrix} 0 & 0 \\ 1 & 0 \end{pmatrix} \begin{pmatrix} y_1(1) \\ y_2(1) \end{pmatrix} = \begin{pmatrix} -\dfrac{1}{\sqrt{\lambda + 1}} \\[2mm] \dfrac{1}{\sqrt{\lambda + 1}} \end{pmatrix}. \qquad (8.17)$$

Our next aim is to determine the associated fundamental matrix $Y(x; -1)$ which satisfies the initial condition $Y(-1; -1) = Y^a = I$. The first column $y_1(x)$ of $Y(x; -1)$ is determined by the initial conditions $y_1(-1) = 1$ and $y_2(-1) = 0$. This IVP can be integrated exactly by the analytical techniques presented in the previous chapters. We get

$$y_1(x) = \begin{pmatrix} -\dfrac{-\lambda^2 + \lambda x^2 + 3\lambda x + x}{\sqrt{x^2 + \lambda}\ \sqrt{(\lambda + 1)^3}} \\[4mm] -\dfrac{\lambda(x + 1)(x^2 - x + 3\lambda + 1)}{\sqrt{(x^2 + \lambda)^3}\ \sqrt{(\lambda + 1)^3}} \end{pmatrix}.$$

The second column $y_2(x)$ of $Y(x; -1)$ is determined by the initial conditions $y_1(-1) = 0$ and $y_2(-1) = 1$. We obtain

$$y_2(x) = \begin{pmatrix} \dfrac{(\lambda - x)(x + 1)}{\sqrt{x^2 + \lambda}\ \sqrt{\lambda + 1}} \\[4mm] -\dfrac{-\lambda^2 + 3\lambda x + \lambda + x^3}{\sqrt{x^2 + \lambda}\ \sqrt{\lambda + 1}} \end{pmatrix}.$$

Thus,

$$Y(x; -1) = \begin{pmatrix} -\dfrac{-\lambda^2 + \lambda x^2 + 3\lambda x + x}{\sqrt{x^2 + \lambda}\ \sqrt{(\lambda + 1)^3}} & \dfrac{(\lambda - x)(x + 1)}{\sqrt{x^2 + \lambda}\ \sqrt{\lambda + 1}} \\[4mm] -\dfrac{\lambda(x + 1)(x^2 - x + 3\lambda + 1)}{\sqrt{(x^2 + \lambda)^3}\ \sqrt{(\lambda + 1)^3}} & -\dfrac{-\lambda^2 + 3\lambda x + \lambda + x^3}{\sqrt{x^2 + \lambda}\ \sqrt{\lambda + 1}} \end{pmatrix}.$$

$$(8.18)$$

It follows

$$Y(1;-1) = \begin{pmatrix} \dfrac{\lambda^2 - 4\lambda - 1}{(\lambda + 1)^2} & \dfrac{2(\lambda - 1)}{\lambda + 1} \\[4mm] -\dfrac{2\lambda(3\lambda + 1)}{(\lambda + 1)^3} & \dfrac{\lambda^2 - 4\lambda - 1}{\lambda + 1} \end{pmatrix}.$$

We are now able to determine the matrix M of the linear algebraic system (8.14):

$$M = B_a + B_b\, Y(1;-1)$$

$$= \begin{pmatrix} 1 & 0 \\ 0 & 0 \end{pmatrix} + \begin{pmatrix} 0 & 0 \\ 1 & 0 \end{pmatrix} \begin{pmatrix} \dfrac{\lambda^2 - 4\lambda - 1}{(\lambda + 1)^2} & \dfrac{2(\lambda - 1)}{\lambda + 1} \\[4mm] -\dfrac{2\lambda(3\lambda + 1)}{(\lambda + 1)^3} & \dfrac{\lambda^2 - 4\lambda - 1}{\lambda + 1} \end{pmatrix}$$

$$= \begin{pmatrix} 1 & 0 \\[4mm] \dfrac{\lambda^2 - 4\lambda - 1}{(\lambda + 1)^2} & \dfrac{2(\lambda - 1)}{\lambda + 1} \end{pmatrix}.$$

Obviously, if $\lambda = 1$, the matrix M is singular and the BVP (8.16) has not a unique solution.

Since the BVP is homogeneous, the particular solution which satisfies $v(-1;-1) = v^a = (0,0)^T$ is $v(x;-1) \equiv (0,0)^T$. Thus, the right-hand side of (8.14) is

$$q = \beta - B_a v(-1;-1) - B_b v(1;-1)$$

$$= \begin{pmatrix} -\dfrac{1}{\sqrt{\lambda + 1}} \\[4mm] \dfrac{1}{\sqrt{\lambda + 1}} \end{pmatrix} - \begin{pmatrix} 1 & 0 \\ 0 & 0 \end{pmatrix} \begin{pmatrix} 0 \\ 0 \end{pmatrix} - \begin{pmatrix} 0 & 0 \\ 1 & 0 \end{pmatrix} \begin{pmatrix} 0 \\ 0 \end{pmatrix}$$

$$= \begin{pmatrix} -\dfrac{1}{\sqrt{\lambda + 1}} \\[4mm] \dfrac{1}{\sqrt{\lambda + 1}} \end{pmatrix}.$$

As can easily be seen, the solution of the algebraic system $M\,c = q$ is

$$c = \begin{pmatrix} -\dfrac{1}{\sqrt{\lambda+1}} \\[2ex] \dfrac{\lambda}{\sqrt{(\lambda+1)^3}} \end{pmatrix}.$$

Finally, the solution of the BVP (8.16) is given by (8.7):

$$y(x) = Y(x; -1)\,c$$

$$= \begin{pmatrix} -\dfrac{-\lambda^2 + \lambda x^2 + 3\lambda x + x}{\sqrt{x^2+\lambda}\,\sqrt{(\lambda+1)^3}} & \dfrac{(\lambda-x)(x+1)}{\sqrt{x^2+\lambda}\,\sqrt{\lambda+1}} \\[3ex] -\dfrac{\lambda(x+1)(x^2-x+3\lambda+1)}{\sqrt{(x^2+\lambda)^3}\,\sqrt{(\lambda+1)^3}} & -\dfrac{-\lambda^2+3\lambda x+\lambda+x^3}{\sqrt{x^2+\lambda}\,\sqrt{\lambda+1}} \end{pmatrix} \begin{pmatrix} -\dfrac{1}{\sqrt{\lambda+1}} \\[2ex] \dfrac{\lambda}{\sqrt{(\lambda+1)^3}} \end{pmatrix}$$

$$= \begin{pmatrix} \dfrac{x}{\sqrt{x^2+\lambda}} \\[2ex] \dfrac{\lambda}{\sqrt{(x^2+\lambda)^3}} \end{pmatrix}. \tag{8.19}$$

Small parameters λ in the BVP (8.16) are interesting for numerical experiments. In that case a layer arises at $x = 0$ as can be seen in Fig. 8.1.

In Table 8.1 some numerical results are presented. We have solved the IVPs (8.12) and (8.13) with the classical Runge-Kutta method of order 4 (see Fig. 7.6). The constant step-size $h = 2/m$ has been chosen such that the following error criterion is satisfied

$$\left\| \begin{pmatrix} \mathrm{err}(-1) \\ \mathrm{err}(0) \\ \mathrm{err}(1) \end{pmatrix} \right\|_{\infty} \le 1e-6,$$

where $\mathrm{err}(t)$ is the difference between the exact and the numerical solution at $x = t$, $t = -1, 0, 1$. In Table 8.1 the column ndgl presents the number of IVPs that had to be integrated.   □

8.3 Method of Complementary Functions

The computational effort of the simple shooting method consists of the numerical integration of the $n + 2$ IVPs (8.12), (8.13), and (8.15) as well as the numerical solution of the system of algebraic equations (8.14). Since, in practice, the

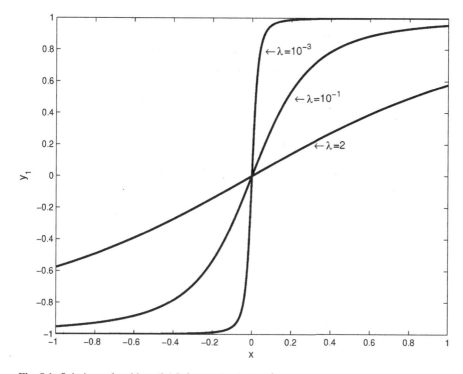

Fig. 8.1 Solutions of problem (8.16) for small values of λ

Table 8.1 Numerical results of the simple shooting method for the BVP (8.16)

λ	m	ndgl	cpu(sec)
$1e-1$	130	1,560	0.07
$1e-2$	780	9,360	0.22
$1e-3$	4,420	53,040	1.1
$1e-4$	24,910	298,920	5.9
$1e-5$	140,200	1,682,400	32.9
$1e-6$	794,000	9,528,000	184.6

dimension n can be a 2- or 3-digit number, it is important to reduce the dimension of the systems of equations which have to be solved. Special attention must be paid to the reduction of the number of IVPs which have to be integrated since the solution of the IVPs requires more than 80–90 % of the total computational effort of a shooting method.

Such a reduction is possible if the boundary conditions are partially separated, i.e.,

$$B_a^{(1)} y(a) = \beta^{(1)},$$

$$B_a^{(2)} y(a) + B_b^{(2)} y(b) = \beta^{(2)}.$$

(8.20)

In that case the boundary matrices B_a and B_b have the form (8.2). The assumption (8.4) implies $\text{rank}(B_a^{(1)}) = p$.

Now, our aim is to reduce the number of fundamental solutions to be integrated (see (8.13)) and the dimension of the algebraic system (8.14) from n to $q = n - p$ by an appropriate choice of the initial matrix Y^a and the initial vector v^a.

Let us start with the system of linear algebraic equations (8.14) of the simple shooting method

$$\left(B_a Y^a + B_b Y^e\right)c = \beta - B_a v^a - B_b v^e,$$

where $Y^e \equiv Y(b; a)$ and $v^e \equiv v(b; a)$. We subdivide $Y(x; a) \in \mathbb{R}^{n \times n}$ and $c \in \mathbb{R}^n$ in accordance with the partitioning (8.2) of the boundary matrices

$$Y(x; a) = \left(W(x; a) | Z(x; a)\right), \quad Y^a = \left(W^a | Z^a\right),$$

$$Y^e = \left(W^e | Z^e\right), \quad c = \begin{pmatrix} w \\ z \end{pmatrix}, \tag{8.21}$$

where $W(x; a), W^a, W^e \in \mathbb{R}^{n \times p}$, $Z(x; a), Z^a, Z^e \in \mathbb{R}^{n \times q}$, $w \in \mathbb{R}^p$, $z \in \mathbb{R}^q$, and $n = p + q$. Substituting (8.21) into the algebraic equations (8.14) gives

$$\begin{pmatrix} B_a^{(1)} W^a & B_a^{(1)} Z^a \\ B_a^{(2)} W^a + B_b^{(2)} W^e & B_a^{(2)} Z^a + B_b^{(2)} Z^e \end{pmatrix} \begin{pmatrix} w \\ z \end{pmatrix}$$

$$= \begin{pmatrix} \beta^{(1)} - B_a^{(1)} v^a \\ \beta^{(2)} - B_a^{(2)} v^a - B_b^{(2)} v^e \end{pmatrix}. \tag{8.22}$$

The $n + 1$ initial vectors of the $n + 1$ IVPs (8.12) and (8.13) which are determined by Y^a and v^a can be freely selected. At this point, let us assume that they satisfy the following two conditions:

$$\text{(i)} \quad B_a^{(1)} Z^a = 0,$$

$$\text{(ii)} \quad B_a^{(1)} v^a = \beta^{(1)}. \tag{8.23}$$

Then, system (8.22) takes on the form

$$\begin{pmatrix} B_a^{(1)} W^a & 0 \\ B_a^{(2)} W^a + B_b^{(2)} W^e & B_a^{(2)} Z^a + B_b^{(2)} Z^e \end{pmatrix} \begin{pmatrix} w \\ z \end{pmatrix}$$

$$= \begin{pmatrix} 0 \\ \beta^{(2)} - B_a^{(2)} v^a - B_b^{(2)} v^e \end{pmatrix}. \tag{8.24}$$

The relations $\mathrm{rank}(B_a^{(1)}) = p$ and $\mathrm{rank}(W^a) = p$ imply that the matrix $B_a^{(1)} W^a \in \mathbb{R}^{p \times p}$ is nonsingular. Thus, from the first block row in (8.24), we get $w = 0$. Using this result in the second block row, we see that the n-dimensional system (8.22) for the vector $c \in \mathbb{R}^n$ is reduced to a system of q linear equations for the reduced vector $z \in \mathbb{R}^q$:

$$\hat{M} z = \hat{q}, \tag{8.25}$$

where

$$\hat{M} \equiv B_a^{(2)} Z^a + B_b^{(2)} Z^e \in \mathbb{R}^{q \times q}, \quad \hat{q} \equiv \beta^{(2)} - B_a^{(2)} v^a - B_b^{(2)} v^e \in \mathbb{R}^q.$$

Now the solution of the BVP with partially separated boundary conditions,

$$\mathcal{L} y(x) = r(x), \quad a \le x \le b,$$

$$B_a^{(1)} y(a) = \beta^{(1)},$$

$$B_a^{(2)} y(a) + B_b^{(2)} y(b) = \beta^{(2)},$$

takes the form

$$y(x) = Z(x; a) z + v(x; a), \tag{8.26}$$

where $Z(x; a) \in \mathbb{R}^{n \times q}$ and $v(x; a) \in \mathbb{R}^n$ are determined by

- q IVPs for the last q columns $Z(x; a)$ of the fundamental matrix $Y(x; a)$:

$$\mathcal{L} Z(x; a) = 0, \quad Z(a; a) = Z^a, \tag{8.27}$$

- One IVP for the particular solution $v(x; a)$:

$$\mathcal{L} v(x; a) = r(x), \quad v(a; a) = v^a. \tag{8.28}$$

Analogous to the simple shooting method, the solution $y(x)$ will not be tabulated on the basis of formula (8.26), but one integrates the associated IVP (8.11). The initial vector $y_a \in \mathbb{R}^n$ is determined by (8.26) as

$$y_a = Z^a z + v^a. \tag{8.29}$$

Now the question arises whether the system of linear equations (8.25) has a unique solution.

Theorem 8.5. *Assume that the BVP* (8.1a) *and* (8.1b) *has a unique solution. If the condition* $B_a^{(1)} Z^a = 0$ *is satisfied, then the system of linear algebraic equations* (8.25) *possesses a unique solution.*

Proof. It is

$$(Y^a)^{-1} Y^a = I_n = \begin{pmatrix} \tilde{W}^a \\ \tilde{Z}^a \end{pmatrix} (W^a | Z^a) = \begin{pmatrix} \tilde{W}^a W^a & \tilde{W}^a Z^a \\ \tilde{Z}^a W^a & \tilde{Z}^a Z^a \end{pmatrix} = \begin{pmatrix} I_p & 0_{p \times q} \\ 0_{q \times p} & I_q \end{pmatrix}.$$

Thus,

$$Y^e (Y^a)^{-1} Z^a = (W^e | Z^e) \begin{pmatrix} \tilde{W}^a \\ \tilde{Z}^a \end{pmatrix} Z^a = (W^e | Z^e) \begin{pmatrix} \tilde{W}^a Z^a \\ \tilde{Z}^a Z^a \end{pmatrix}$$

$$= (Y^e | Z^e) \begin{pmatrix} 0_{p \times q} \\ I_q \end{pmatrix} = Z^e.$$

We now form the expression

$$M (Y^a)^{-1} Z^a = (B_a Y^a + B_b Y^e) (Y^a)^{-1} Z^a = B_a Z^a + B_b Z^e$$

$$= \begin{pmatrix} B_a^{(1)} Z^a \\ \hat{M} \end{pmatrix}.$$

The matrix on the left-hand side of this expression has the rank q since $\text{rank}(M) = n$, $\text{rank}((Y^a)^{-1}) = n$, and $\text{rank}(Z^a) = q$. The matrix of the right-hand side must have the rank q, too. The condition (i) in (8.23) implies that the matrix $\hat{M} \in \mathbb{R}^{q \times q}$ must have the rank q, i.e., the system matrix $\hat{M}$ is nonsingular. Thus, system (8.25) can be solved uniquely. ∎

Now, we will determine the initial matrix Z^a and the initial vector v^a such that the conditions (8.23) are fulfilled. For this purpose, a QR-factorization of the matrix $(B_a^{(1)})^T \in \mathbb{R}^{n \times p}$ is computed:

$$(B_a^{(1)})^T = Q \begin{pmatrix} U \\ 0 \end{pmatrix}, \tag{8.30}$$

where $Q \in \mathbb{R}^{n \times n}$ is orthogonal and $U \in \mathbb{R}^{p \times p}$ is an upper triangular matrix. Let us subdivide Q as follows:

$$Q = (Q_1 | Q_2), \quad Q_1 \in \mathbb{R}^{n \times p} \quad \text{and} \quad Q_2 \in \mathbb{R}^{n \times q}. \tag{8.31}$$

Substituting this into (8.30) gives

$$(B_a^{(1)})^T = Q \begin{pmatrix} U \\ 0 \end{pmatrix} = (Q_1 | Q_2) \begin{pmatrix} U \\ 0 \end{pmatrix} = Q_1 U.$$

The (numerically stable) computation of the matrices Q and U is realized by a sequence of p Householder transformations. If we set

$$Z^a \equiv Q_2 \quad \text{and} \quad v^a \equiv Q_1 U^{-T} \beta^{(1)}, \tag{8.32}$$

the conditions (8.23) are satisfied. This can be shown as follows:

(i) $B_a^{(1)} Z^a = U^T \underbrace{Q_1^T Q_2}_{=0} = 0$

(ii) $B_a^{(1)} v^a = U^T \underbrace{\underbrace{Q_1^T Q_1}_{I} U^{-T} \beta^{(1)}}_{I} = \beta^{(1)}.$

The method described above is called *method of complementary functions*. The name refers to the q columns of the complete fundamental system $Y(x; a)$ which are contained in the reduced matrix $Z(x; a)$. These columns are also denoted as *complementary functions*. The computational effort of the method of complementary functions consists essentially of the integration of the $q + 2$ IVPs (8.11), (8.27) and (8.28), the solution of the q-dimensional system of linear equations (8.25) as well as the p Householder transformations to generate the factorization (8.30). In comparison with the simple shooting method the number of IVPs which have to be integrated is reduced by p. However, the algebraic work is more or less the same.

Example 8.6. Given the BVP

$$\mathscr{L}y(x) = r(x), \quad a \le x \le b,$$

$$\mathscr{B}y(x) = \beta,$$

with $n = 2$, $q = 1$ and the boundary matrices

$$B_a = \begin{pmatrix} 1 & 0 \\ a_1 & a_2 \end{pmatrix}, \quad B_b = \begin{pmatrix} 0 & 0 \\ b_1 & b_2 \end{pmatrix}.$$

Obviously, the boundary conditions are partially separated, and it holds

$$B_a^{(1)} = (1, 0), \quad B_a^{(2)} = (a_1, a_2), \quad B_b^{(2)} = (b_1, b_2).$$

The matrices Q and U of the factorization (8.30) are

$$Q_1 = (1,0)^T, \quad Q_2 = (0,1)^T, \quad U = (1).$$

$\square$

It is also possible to determine a matrix Z^a and a vector v^a which satisfy the conditions (8.23) by an elimination method. Here, the following LU-factorization of the matrix $B_a^{(1)}$ with a stabilizing row pivoting is computed

$$B_a^{(1)} = (L_1 \,|\, 0) \, U \, P = (L_1 \,|\, 0) \begin{pmatrix} U_1 & H \\ 0 & I_q \end{pmatrix} P,$$

where $L_1 \in \mathbb{R}^{p \times p}$ is a lower triangular matrix, $U \in \mathbb{R}^{n \times n}$ is an upper triangular matrix whose diagonal elements are equal to one and the absolute value of the other elements is smaller or equal to one, $I_q \in \mathbb{R}^{q \times q}$ is the q-dimensional unit matrix, and $P \in \mathbb{R}^{n \times n}$ is a permutation matrix describing the exchange of the columns during the pivoting process. If we set

$$Z^a \equiv P^T \begin{pmatrix} -U_1^{-1} H \\ I_q \end{pmatrix} \quad \text{and} \quad v^a \equiv P^T \begin{pmatrix} U_1^{-1} L_1^{-1} \beta^{(1)} \\ 0 \end{pmatrix}, \tag{8.33}$$

then the conditions (8.23) are clearly satisfied.

Example 8.7. Let us consider the BVP (8.16) of Example 8.4. The boundary conditions are partially separated, and therefore, we will use the method of complementary functions to solve this BVP.

At first we have to compute the QR-factorization (8.30), i.e.,

$$(B_a^{(1)})^T = Q \begin{pmatrix} U \\ 0 \end{pmatrix} = (Q_1 | Q_2) \begin{pmatrix} U \\ 0 \end{pmatrix} = \begin{pmatrix} 1 & 0 \\ 0 & 1 \end{pmatrix} \begin{pmatrix} 1 \\ 0 \end{pmatrix}. \tag{8.34}$$

Thus,

$$Q_1 = \begin{pmatrix} 1 \\ 0 \end{pmatrix}, \quad Q_2 = \begin{pmatrix} 0 \\ 1 \end{pmatrix}, \quad U = (1).$$

The initial vectors for the IVPs (8.27) and (8.28) are determined by (8.32):

$$Z^a = \begin{pmatrix} 0 \\ 1 \end{pmatrix}, \quad v^a = \begin{pmatrix} 1 \\ 0 \end{pmatrix} \cdot 1 \cdot \left(-\frac{1}{\sqrt{\lambda + 1}} \right) = -\begin{pmatrix} \frac{1}{\sqrt{\lambda + 1}} \\ 0 \end{pmatrix}.$$

Using the analytical techniques described in the previous chapters, we calculate the solutions of the IVPs (8.27) and (8.28):

$$
Z(x; -1) = \begin{pmatrix} \dfrac{(\lambda - x)(x + 1)}{\sqrt{x^2 + \lambda}\,\sqrt{\lambda + 1}} \\[4mm] -\dfrac{-\lambda^2 + 3\lambda x + \lambda + x^3}{\sqrt{x^2 + \lambda}\,\sqrt{\lambda + 1}} \end{pmatrix},
$$

$$
v(x; -1) = \begin{pmatrix} \dfrac{-\lambda^2 + \lambda x^2 + 3\lambda x + x}{\sqrt{x^2 + \lambda}\,(\lambda + 1)^2} \\[4mm] \dfrac{\lambda(x + 1)(x^2 - x + 3\lambda + 1)}{\sqrt{(x^2 + \lambda)^3}\,(\lambda + 1)^2} \end{pmatrix}. \tag{8.35}
$$

Now, the matrix $\hat{M}$ of the system of linear equations (8.25) is

$$
\hat{M} = (0, 0)\begin{pmatrix} 0 \\ 1 \end{pmatrix} + (1, 0)\begin{pmatrix} \dfrac{2(\lambda - 1)}{\lambda + 1} \\[3mm] \dfrac{\lambda^2 - 4\lambda - 1}{\lambda + 1} \end{pmatrix} = \dfrac{2(\lambda - 1)}{\lambda + 1}.
$$

The corresponding right-hand side $\hat{q}$ is

$$
\hat{q} = \dfrac{1}{\sqrt{\lambda + 1}} + (0, 0)\begin{pmatrix} \dfrac{1}{\sqrt{\lambda + 1}} \\[2mm] 0 \end{pmatrix} - (1, 0)\begin{pmatrix} \dfrac{-\lambda^2 + 4\lambda + 1}{\sqrt{\lambda + 1}(\lambda + 1)^2} \\[3mm] \dfrac{2\lambda(3\lambda + 1)}{\sqrt{(\lambda + 1)^3}(\lambda + 1)} \end{pmatrix}
$$

$$
= \dfrac{1}{\sqrt{\lambda + 1}} - \dfrac{-\lambda^2 + 4\lambda + 1}{\sqrt{\lambda + 1}(\lambda + 1)^2} = \dfrac{2\lambda(\lambda - 1)}{\sqrt{\lambda + 1}(\lambda + 1)^2}.
$$

Thus, the system of linear equations (8.25) is a single equation

$$
\dfrac{2(\lambda - 1)}{\lambda + 1}\, z = \dfrac{2\lambda(\lambda - 1)}{\sqrt{\lambda + 1}(\lambda + 1)^2}
$$

with the solution

$$
z = \dfrac{\lambda}{\sqrt{\lambda + 1}(\lambda + 1)}.
$$

Table 8.2 Numerical results
of the method of
complementary functions for
the BVP (8.16)

λ	m	ndgl	cpu
$1e-1$	130	1,040	0.06
$1e-2$	776	6,208	0.16
$1e-3$	4,420	35,360	0.73
$1e-4$	24,902	199,216	4.0
$1e-5$	140,200	1,121,600	22.2
$1e-6$	791,000	6,328,000	124.6

Using formula (8.26) the solution of the BVP (8.16) can be represented as

$$y(x) = Z(x;-1)z + v(x;-1)$$

$$= \begin{pmatrix} \dfrac{(\lambda - x)(x+1)}{\sqrt{x^2+\lambda}\,\sqrt{\lambda+1}} \\[2ex] -\dfrac{-\lambda^2 + 3\lambda x + \lambda + x^3}{\sqrt{x^2+\lambda}\,\sqrt{\lambda+1}} \end{pmatrix} \dfrac{\lambda}{\sqrt{\lambda+1}(\lambda+1)} + \begin{pmatrix} \dfrac{-\lambda^2 + \lambda x^2 + 3\lambda x + x}{\sqrt{x^2+\lambda}\,(\lambda+1)^2} \\[2ex] \dfrac{\lambda(x+1)(x^2 - x + 3\lambda + 1)}{\sqrt{(x^2+\lambda)^3}\,(\lambda+1)^2} \end{pmatrix}$$

$$= \begin{pmatrix} \dfrac{x}{\sqrt{x^2+\lambda}} \\[2ex] \dfrac{\lambda}{\sqrt{(x^2+\lambda)^3}} \end{pmatrix} . \tag{8.36}$$

In Table 8.2 we give the numerical results of the method of complementary
functions for the same parameter values presented in Table 8.1. The numerical
solution of the IVPs has been realized in the same way as described in Example 8.4.
A direct comparison shows that the method of complementary functions is really
more effective than the simple shooting method.   □

8.4   Stability of Simple Shooting

As mentioned before, the idea of the simple shooting method is based on the fact
that for each x the columns of the fundamental matrix $Y(x;a)$ form a basis for the
solution manifold of the homogeneous ODEs. Theoretically, these columns remain
independent forever, although the angles between them may become very small. In
practice, the columns become spoiled by truncation and roundoff errors, the stronger
as $|x - a|$ grows larger or as the differences between the eigenvalues of the matrix
$A(x)$ are bigger. Therefore, the system of linear algebraic equations (8.14) may be
perturbed so strongly that the approximation of the solution may become very poor.
Especially this comes true if the ODEs (8.1a) are instable in the following sense.

Definition 8.8 (See [11]). The system of differential equations (8.1a) is called (ε, K)-instable if $y(x)$ is a solution and there exists another one $\hat{y}(x)$ with

$$\|y(a) - \hat{y}(a)\| \leq \varepsilon, \quad \text{and}$$

$$\|y(x^*) - \hat{y}(x^*)\| \geq K,$$

(8.37)

for at least one $x^* \in [a, b]$. □

In (8.37) ε and K denote a small and a large, respectively, positive real number. In particular, this means a blowup of the trajectories of the associated IVPs.

The occurrence of (ε, K)-instable ODEs has at least two consequences for simple shooting codes:

- The integration method fails (the step-size h tends to zero and the right boundary cannot be reached).
- The matrix M is ill-conditioned (the system (8.14) depends very sensitively on the problem data).

Let us demonstrate the (ε, K)-instability by an example.

Example 8.9. Given the following BVP (see [16]),

$$y''(x) = \lambda^2 (y(x) + \cos^2(\pi x)) + 2\pi^2 \cos(2\pi x),$$

$$y(0) = 0, \quad y(1) = 0,$$

(8.38)

where λ is a real parameter.

The formulation of (8.38) as a first-order system is

$$y_1'(x) = y_2(x), \quad y_2'(x) = \lambda^2 (y_1(x) + \cos^2(\pi x)) + 2\pi \cos(2\pi x),$$

$$\begin{pmatrix} 1 & 0 \\ 0 & 0 \end{pmatrix} \begin{pmatrix} y_1(0) \\ y_2(0) \end{pmatrix} + \begin{pmatrix} 0 & 0 \\ 1 & 0 \end{pmatrix} \begin{pmatrix} y_1(1) \\ y_2(1) \end{pmatrix} = \begin{pmatrix} 0 \\ 0 \end{pmatrix}.$$

(8.39)

The corresponding fundamental matrix $Y(x; 0)$ which satisfies $Y(0; 0) = I$ is

$$Y(x; 0) = \begin{pmatrix} \cosh(\lambda x) & \sinh(\lambda x)/\lambda \\ \lambda \sinh(\lambda x) & \cosh(\lambda x) \end{pmatrix}.$$

(8.40)

The particular solution $v(x; 0)$ with $v(0; 0) = (0, 0)^T$ is

$$v(x; 0) = \begin{pmatrix} \sin^2(\pi x) + 2\sinh^2((\lambda x)/2) \\ \pi \sin(2\pi x) + \lambda \sinh(\lambda x) \end{pmatrix}.$$

(8.41)

Table 8.3 Condition numbers of M

λ	5	10	15	20	25
cond(M)	$3.9\,e2$	$1.1\,e5$	$2.5\,e7$	$4.9\,e9$	$9.0\,e11$

Thus, the matrix M of the linear algebraic system (8.14) is

$$M = \begin{pmatrix} 1 & 0 \\ \cosh\lambda & \sinh(\lambda)/\lambda \end{pmatrix}. \tag{8.42}$$

In Table 8.3 the condition numbers $\text{cond}_2(M) = \|M\|_2\|M^{-1}\|_2$ of the system matrix M for increasing values of λ are presented.

It shows that problem (8.6) is highly (ε, K)-instable.   □

8.5 Multiple Shooting Method

As we have seen in the previous section, the simple shooting techniques are doomed to fail if the interval length becomes too large. An obvious generalization is then to use it as the basic algorithm in multiple shooting.

In fact, the multiple shooting method is constructed to overcome the mentioned difficulties of the simple shooting method. It is a segmentation technique and a prototype of a parallel algorithm. We think that the first paper dealing with segmentation techniques for shooting methods is that of Morrison et al. [10]. The aim of the authors was to combine the advantages of shooting with the advantages of finite difference methods.

The principle of the multiple shooting method can be described as follows. The interval $[a, b]$ is *suitably* subdivided by $m + 1$ shooting or segmentation points τ_j,

$$a = \tau_0 < \tau_1 < \tau_2 < \cdots < \tau_m = b. \tag{8.43}$$

On each segment $[\tau_j, \tau_{j+1}]$, $j = 0, \ldots, m - 1$, the solution of the ODE can be represented by the principle of superposition in the form

$$y(x) = y_j(x) = Y(x; \tau_j)c_j + v(x; \tau_j), \quad x \in [\tau_j, \tau_{j+1}]. \tag{8.44}$$

The fundamental matrix $Y(x; \tau_j)$ is the solution of the matrix-IVP

$$\mathscr{L}Y(x; \tau_j) = 0, \quad \tau_j \le x \le \tau_{j+1},$$

$$Y(\tau_j; \tau_j) = Y_j^a, \quad Y_j^a \in \mathbb{R}^{n \times n} \text{ with } \det(Y_j^a) \ne 0. \tag{8.45}$$

The particular solution $v(x; \tau_j)$ is defined by the IVP

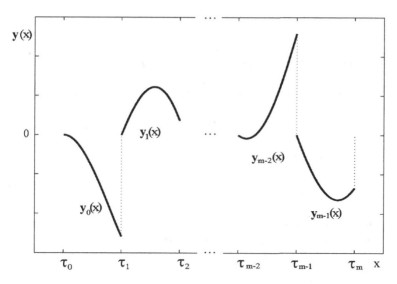

Fig. 8.2 Principle of multiple shooting

$$\mathcal{L}v(x;\tau_j) = r(x), \quad \tau_j \le x \le \tau_{j+1},$$
$$v(\tau_j;\tau_j) = v_j^a, \quad v_j^a \in \mathbb{R}^n. \tag{8.46}$$

Now, the solution pieces $y_j(x)$, $j = 0,\ldots,m-1$, are combined by appropriate choice of the $m-1$ vectors $c_j \in \mathbb{R}^n$ such that a continuous solution of the ODE results which also satisfies the boundary conditions (see Fig. 8.2). Let us write the continuity conditions at the interior nodes first:

$$y_0(\tau_1) = y_1(\tau_1),$$
$$y_1(\tau_2) = y_2(\tau_2),$$
$$\vdots = \vdots \tag{8.47}$$
$$y_{m-2}(\tau_{m-1}) = y_{m-1}(\tau_{m-1}).$$

Using the representation (8.44) of $y_j(x)$, these equations can be written in the form

$$Y_j^e c_j + v_j^e = Y_{j+1}^a c_{j+1} + v_{j+1}^a, \quad j = 0,\ldots,m-2,$$

where we have used the abbreviations

$$Y_j^e \equiv Y(\tau_{j+1};\tau_j), \quad v_j^e \equiv v(\tau_{j+1};\tau_j).$$

After a rearrangement of the terms, we obtain

$$c_{j+1} - \left(Y_{j+1}^a\right)^{-1} Y_j^e c_j = \left(Y_{j+1}^a\right)^{-1} \left(v_j^e - v_{j+1}^a\right), \quad j = 0, \dots, m-2. \quad (8.48)$$

Finally, the ansatz (8.44) is substituted into the boundary conditions:

$$B_a[Y(\tau_0; \tau_0)c_0 + v(\tau_0; \tau_0)] + B_b[Y(\tau_m; \tau_{m-1})c_{m-1} + v(\tau_m; \tau_{m-1})] = \beta.$$

Rearranging the terms yields

$$B_a Y_0^a c_0 + B_b Y_{m-1}^e c_{m-1} = \beta - B_a v_0^a - B_b v_{m-1}^e. \quad (8.49)$$

If we combine the continuity conditions (8.48) and the boundary conditions (8.49), we get the mn-dimensional system of linear algebraic equations

$$M^{(m)} c^{(m)} = q^{(m)}, \quad (8.50)$$

where

$$M^{(m)} \equiv \begin{pmatrix} -(Y_1^a)^{-1} Y_0^e & I & & & \\ & -(Y_2^a)^{-1} Y_1^e & I & & \\ & & \ddots & \ddots & \\ & & & -(Y_{m-1}^a)^{-1} Y_{m-2}^e & I \\ B_a Y_0^a & & & & B_b Y_{m-1}^e \end{pmatrix},$$

$$q^{(m)} \equiv \begin{pmatrix} (Y_1^a)^{-1}(v_0^e - v_1^a) \\ \vdots \\ (Y_{m-1}^a)^{-1}(v_{m-2}^e - v_{m-1}^a) \\ \beta - B_a v_0^a - B_b v_{m-1}^e \end{pmatrix} \quad \text{and} \quad c^{(m)} \equiv \begin{pmatrix} c_0 \\ c_1 \\ \vdots \\ c_{m-1} \end{pmatrix}.$$

If the vectors $c_j \in \mathbb{R}^n$ are determined from system (8.50), then on each segment $[\tau_j, \tau_{j+1}]$, an associated initial vector can be computed by formula (8.44) as

$$y(\tau_j) = Y_j^a c_j + v_j^a \equiv s_j. \quad (8.51)$$

Now, it is possible to compute an approximation of the solution of the BVP (8.1a) and (8.1b) at an arbitrary $\hat{x} \in [\tau_j, \tau_{j+1}]$ by integrating the IVP

$$\mathscr{L} y(x) = r(x), \quad \tau_j \le x \le \tau_{j+1},$$

$$y(\tau_j) = s_j \quad (8.52)$$

from τ_j to $\hat{x}$.

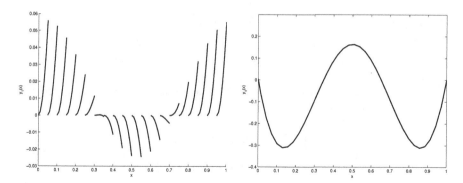

Fig. 8.3 Multiple shooting solution of problem (8.38) with $\lambda = 5$ and $m = 20$

If the integrations (8.45) and (8.46) are realized with the special initial conditions

$$Y_j^a = I \in \mathbb{R}^{n \times n} \quad \text{and} \quad v_j^a = \mathbf{0} \in \mathbb{R}^n, \tag{8.53}$$

we have the so-called *standard form* of the multiple shooting method. In that case the inverting of Y_{j+1}^a can be dropped.

In Fig. 8.3 we show the initial trajectories and the numerical result of the standard form of the multiple shooting method for problem (8.38).

Now, the question arises which conditions guarantee the regularity of the matrix $M^{(m)}$ such that the system of linear algebraic equations (8.50) has a unique solution. The answer gives the following theorem.

Theorem 8.10. *The matrix $M^{(m)}$ is nonsingular if and only if the matrix M of the simple shooting method (see formula (8.14)) is nonsingular.*

Proof. See, e.g., [5]. ∎

8.6   Stability of Multiple Shooting

The use of segmentation techniques in shooting methods eliminates the first disadvantage of the simple shooting method. Namely, the shooting points τ_j can always be chosen (automatically) such that

$$\|Y_j^e\| \leq 1 + \gamma, \quad 0 \leq j \leq m - 1, \quad \gamma \geq 0,$$

is fulfilled. Therefore, the integration methods work well and $\|M^{(m)}\|$ can be controlled.

With regard to an estimation of the condition number $\text{cond}(M^{(m)})$, it remains to study the behavior of $\|(M^{(m)})^{-1}\|$. It can be shown (see, e.g., [5]) that

$$(M^{(m)})^{-1} = \begin{pmatrix} G(\tau_0, \tau_1) & \cdots & G(\tau_0, \tau_{m-1}) & Y(\tau_0, \tau_0)M^{-1} \\ \vdots & & \vdots & \vdots \\ G(\tau_{m-1}, \tau_1) & \cdots & G(\tau_{m-1}, \tau_{m-1}) & Y(\tau_{m-1}, \tau_0)M^{-1} \end{pmatrix}, \qquad (8.54)$$

where M is the system matrix of the simple shooting method and $G(x, \tau)$ denotes the Green's function which is defined as

$$G(x, \tau) \equiv \begin{cases} Y(x;a)B_a Y(a;a)Y(\tau;a)^{-1}, \ x \geq \tau \\ -Y(x;a)B_b Y(b;a)Y(\tau;a)^{-1}, \ x < \tau. \end{cases} \qquad (8.55)$$

Formula (8.54) shows that the block matrix elements of $(M^{(m)})^{-1}$ are only functions of their position and do not depend explicitly on the segmentation. In particular, they are uniformly bounded such that there is an estimation

$$\|(M^{(m)})^{-1}\| \leq Lm, \qquad (8.56)$$

where L is a real constant. But it may be that L has a value of the order of magnitude of the strongly increasing solutions of the ODEs.

In the following theorem, a condition is given under which the condition number $\text{cond}(M^{(m)})$ has a moderate size.

Theorem 8.11. *Assume that the solution* $y(x)$ *of the BVP* (8.1a) *and* (8.1b) *depends continuously on the data of the problem, i.e., the BVP is well conditioned. This dependence can be formulated as follows (see [5]):*

$$\|y(x)\|_{C[a,b]^n} \leq \max \left\{ \kappa_1 \|r(x)\|_{C[a,b]^n}, \kappa_2 \|\beta\|_{\mathbb{R}^n} \right\}. \qquad (8.57)$$

Then, the condition number $\text{cond}(M^{(m)})$ *grows at most linearly in m and does not produce any numerical difficulties.*

Proof. See, e.g., [5] or [13]. ∎

In the next example, we demonstrate that the estimation (8.56) is a worst-case estimation.

Example 8.12. Given the BVP

$$u''(x) = \lambda^2 u(x) + f(x, \lambda), \quad x \in [0, 1],$$

$$u(0) = u_0, \quad u(1) = u_1. \qquad (8.58)$$

For large values of λ, this BVP is the prototype of a highly (ε, K)-instable problem. It can be used as a stability equation in analogy to the equation $y'(x) = \lambda y(x)$ which is used in the theory of absolute stable IVPs (see Chap. 7).

We transform problem (8.58) into a first-order system (8.1a) and (8.1b) by setting

$$y(x) \equiv \begin{pmatrix} u(x) \\ u'(x) \end{pmatrix}, \quad A(x) \equiv \begin{pmatrix} 0 & 1 \\ \lambda^2 & 0 \end{pmatrix}, \quad r(x,\lambda) \equiv \begin{pmatrix} 0 \\ f(x,\lambda) \end{pmatrix},$$

$$B_a \equiv \begin{pmatrix} 1 & 0 \\ 0 & 0 \end{pmatrix}, \quad B_b \equiv \begin{pmatrix} 0 & 0 \\ 1 & 0 \end{pmatrix}, \quad \beta \equiv \begin{pmatrix} u_0 \\ u_1 \end{pmatrix}.$$

For an equidistant grid of shooting points $\tau_k = k/m$, $k = 0, \ldots, m$, the fundamental matrices are

$$Y(\tau_{k+1}; \tau_k) = \begin{pmatrix} \text{ch} & \text{sh}/\lambda \\ \lambda\,\text{sh} & \text{ch} \end{pmatrix},$$

with $\text{ch} \equiv \cosh(\lambda/m)$ and $\text{sh} \equiv \sinh(\lambda/m)$. Thus, the matrix $M^{(m)}$ of the linear algebraic system (8.50) can be written in the form

$$\begin{pmatrix}
-\begin{pmatrix} \text{ch} & \text{sh}/\lambda \\ \lambda\,\text{sh} & \text{ch} \end{pmatrix} & \begin{pmatrix} 1 & 0 \\ 0 & 1 \end{pmatrix} & & & \\
& -\begin{pmatrix} \text{ch} & \text{sh}/\lambda \\ \lambda\,\text{sh} & \text{ch} \end{pmatrix} & \begin{pmatrix} 1 & 0 \\ 0 & 1 \end{pmatrix} & & \\
& & \ddots & \ddots & \\
& & & -\begin{pmatrix} \text{ch} & \text{sh}/\lambda \\ \lambda\,\text{sh} & \text{ch} \end{pmatrix} & \begin{pmatrix} 1 & 0 \\ 0 & 1 \end{pmatrix} \\
\begin{pmatrix} 1 & 0 \\ 0 & 0 \end{pmatrix} & & & & \begin{pmatrix} 0 & 0 \\ \text{ch} & \text{sh}/\lambda \end{pmatrix}
\end{pmatrix}.$$

It is not difficult to show that cond($M^{(m)}$) grows at least as fast as $e^{\lambda/m}$ for $\lambda/m \to \infty$. In particular, this implies:

- For a fixed value of m (i.e., the segmentation is not changed) but an increasing λ, the condition of the algebraic system (8.50) grows exponentially.
- For a fixed value of λ but a increasing m (i.e., the segmentation is refined), the condition of the algebraic system (8.50) is continually improved.

A multiple shooting algorithm is called *E-stable* if it reflects the second feature.

Obviously, problem (8.38) is a special case of (8.58). A look at Table 8.3 shows that for $\lambda = 5$ the condition number of the matrix M of the simple shooting method is $3.9\,e2$. For larger values of λ, the condition number cond(M) is growing strongly. In Table 8.4 we demonstrate that by increasing the number of shooting points, the condition number of the multiple shooting method cond($M^{(m)}$) can be bounded.

$\square$

Table 8.4 Condition numbers of $M^{(m)}$

λ	5	10	15	20	25
m	1	4	7	13	25
$\text{cond}(M^{(m)})$	3.9 e2	6.2 e2	9.8 e2	9.8 e2	9.7 e2

8.7 Numerical Treatment of the Linear Algebraic System

In this section we look for numerical methods by which the linear algebraic system (8.50) can be solved. In particular, we aim at ensuring that the resulting multiple shooting algorithm is E-stable.

A frequently used numerical method is the compactification or condensing method (see [2, 5, 16]). It is based on the following factorization and requires less storage as all other methods:

$$
M^{(m)} = \begin{pmatrix} I & & \\ & \ddots & \\ & & I \\ Q_1 \; Q_2 \; \cdots \; Q_m \end{pmatrix} \begin{pmatrix} I & & \\ -Y_0^e & I & \\ & \ddots & \ddots \\ & & -Y_{m-2}^e \; I \end{pmatrix}.
$$

At first the (small) n-dimensional system

$$
\left(B_a + B_b \prod_{j=0}^{m-1} Y_{m-1-j}^e \right) c_0 = \beta - B_b v_{m-1}^e - B_b \sum_{j=0}^{m-2} \left(\prod_{l=0}^{m-2-j} Y_{m-1-l}^e \right) v_j^e \quad (8.59)
$$

is solved. Then, the other n-dimensional block components $c_1, \ldots, c_{m-1}$ of the solution vector $c^{(m)} \in \mathbb{R}^{mn}$ are computed by the recurrence formula

$$
c_{j+1} = Y_j^e c_j + v_j^e, \quad j = 0, \ldots, m-2. \quad (8.60)
$$

However, the corresponding multiple shooting method is neither E-stable nor stable in the usual sense. The reason is that in exact arithmetics the system matrix in (8.59) can be rewritten as

$$
B_a + B_b Y_{m-1}^e Y_{m-2}^e \cdots Y_0^e = B_a + B_b Y(b; a) = M
$$

since the fundamental matrices are Wronskian matrices satisfying

$$
Y(x; t) Y(t; a) = Y(x; a).
$$

Thus, system (8.59) has the same system matrix M as the simple shooting method. The addition of further shooting points has no influence on the condition of system (8.59) although the condition of $M^{(m)}$ is improved. Moreover, in [3] it is shown that the recurrence formula (8.60) is numerically instable. Therefore, multiple shooting algorithms on the basis of this compactification method should be used only for very well-conditioned BVPs.

The numerical analysis of linear algebraic equations shows that Gaussian elimination with partial pivoting and scaling is an in-detail-studied and stable algorithm. Its application to system (8.50) results in an E-stable multiple shooting method. To eliminate effects of improper scaling, some steps of iterative refinement (at least one) are necessary (see, e.g., [15]). In order to take advantage of the sparsity of the matrix $M^{(m)}$, the linear system solver has to be formulated in a packed form of storage.

There are some other stable numerical elimination techniques like the block LU- and QR-factorizations presented in [18–20] and the stabilizing transformation published in [4, 5].

8.8 Stabilized March Method

If the IVPs (8.27) and (8.28) are unstable and/or the matrix $\hat{M}$ of the linear system (8.25) is ill-conditioned, then a stabilization of the method of complementary functions can be achieved by segmentation techniques analogously to the transition from simple to multiple shooting.

Before we describe the method, let us have a look at the numerical effort of the shooting techniques studied in the previous sections which is summarized in Table 8.5. The table contains the following important question: Is it possible to reduce the number of IVPs to be integrated as well as the dimension of the system of linear algebraic equations as this was true for the transition from the simple shooting method to the method of complementary functions? Moreover, if this is the case, are the data given in Table 8.5 valid? To answer these questions, let us derive the stabilized march method from the system of linear algebraic equations of the multiple shooting method.

Table 8.5 Numerical effort of the shooting methods

	Simple shooting	Method of complementary functions
Integrations	$n + 1$	$q + 1$
Dimension of the algebraic system	n	q
	Multiple shooting	Stabilized march method
Integrations	$m(n + 1)$	? $(q(n + 1))$?
Dimension of the algebraic system	mn	? (qn) ?

In accordance with the partitioning (8.2) of the boundary matrices B_a and B_b, we write the ansatz (8.44) in the form

$$y(x) = y_j(x) = \left(W(x; \tau_j) \mid Z(x; \tau_j)\right) \begin{pmatrix} w_j \\ z_j \end{pmatrix} + v(t; \tau_j), \quad x \in [\tau_j, \tau_{j+1}], \quad (8.61)$$

where $W(x; \tau_j) \in \mathbb{R}^{n \times p}$, $Z(x; \tau_j) \in \mathbb{R}^{n \times q}$, $w_j \in \mathbb{R}^p$ and $z_j \in \mathbb{R}^q$. Substituting (8.60) into the boundary conditions

$$\begin{pmatrix} B_a^{(1)} \\ B_a^{(2)} \end{pmatrix} y(a) + \begin{pmatrix} 0 \\ B_b^{(2)} \end{pmatrix} y(b) = \begin{pmatrix} \beta^{(1)} \\ \beta^{(2)} \end{pmatrix},$$

we obtain

$$\begin{pmatrix} B_a^{(1)} \\ B_a^{(2)} \end{pmatrix} (W_0^a \mid Z_0^a) \begin{pmatrix} w_0 \\ z_0 \end{pmatrix} + \begin{pmatrix} 0 \\ B_b^{(2)} \end{pmatrix} (W_{m-1}^e \mid Z_{m-1}^e) \begin{pmatrix} w_{m-1} \\ z_{m-1} \end{pmatrix}$$

$$= \begin{pmatrix} \beta^{(1)} \\ \beta^{(2)} \end{pmatrix} - \begin{pmatrix} B_a^{(1)} \\ B_a^{(2)} \end{pmatrix} v_0^a - \begin{pmatrix} 0 \\ B_b^{(2)} \end{pmatrix} v_{m-1}^e. \quad (8.62)$$

The initial vector v_0^a and the initial matrix Z_0^a can be freely selected. Therefore, we assume that they satisfy the following two conditions:

$$\text{(i)} \quad B_a^{(1)} Z_0^a = 0,$$
$$\quad (8.63)$$
$$\text{(ii)} \quad B_a^{(1)} v_0^a = \beta^{(1)}.$$

Since $\text{rank}(B_a^{(1)}) = p$ and $W_0^a \in \mathbb{R}^{n \times p}$ consists of the first p columns of the fundamental matrix Y_0^a, the matrix $B_a^{(1)} W_0^a \in \mathbb{R}^{p \times p}$ has full rank. Thus, from the first block row of (8.62), we get $w_0 = 0$. As we will see later, it holds $w_{m-1} = 0$. This implies that the second block row of (8.62) is reduced to

$$B_a^{(2)} Z_0^a z_0 + B_b^{(2)} Z_{m-1}^e z_{m-1} = \beta^{(2)} - B_a^{(2)} v_0^a - B_b^{(2)} v_{m-1}^e. \quad (8.64)$$

Note that the conditions (8.63) agree with Eqs. (8.23) for the method of complementary functions.

On the first segment $[\tau_0, \tau_1]$, the corresponding solution $y_0(x)$ of the BVP (8.1a), (8.1b), (8.2) can be represented in the form

$$y_0(x) = Z(x; \tau_0) z_0 + v(x; \tau_0). \quad (8.65)$$

Now, we consider the continuity conditions at the inner shooting points τ_j:

$$(W_{j-1}^e | Z_{j-1}^e) \begin{pmatrix} w_{j-1} \\ z_{j-1} \end{pmatrix} + v_{j-1}^e = (W_j^a | Z_j^a) \begin{pmatrix} w_j \\ z_j \end{pmatrix} + v_j^a, \quad j = 1, \ldots, m-1. \quad (8.66)$$

To simplify these equations, let us assume that $w_{j-1} = 0$ is already valid. Then,

$$Z_{j-1}^e z_{j-1} - (W_j^a | Z_j^a) \begin{pmatrix} w_j \\ z_j \end{pmatrix} = v_j^a - v_{j-1}^e. \quad (8.67)$$

We set

$$\begin{pmatrix} \tilde{W}_j^a \\ \tilde{Z}_j^a \end{pmatrix} \equiv (W_j^a | Z_j^a)^{-1}, \quad \tilde{W}_j^a \in \mathbb{R}^{p \times n}, \ \tilde{Z}_j^a \in \mathbb{R}^{q \times n}.$$

Multiplying (8.67) by this matrix yields

$$\begin{pmatrix} \tilde{W}_j^a \\ \tilde{Z}_j^a \end{pmatrix} Z_{j-1}^e z_{j-1} - \begin{pmatrix} w_j \\ z_j \end{pmatrix} = \begin{pmatrix} \tilde{W}_j^a \\ \tilde{Z}_j^a \end{pmatrix} (v_j^a - v_{j-1}^e). \quad (8.68)$$

So that $w_j = 0$ also holds, it must be requested that

$$\begin{array}{ll} \text{(i)} & \tilde{W}_j^a Z_{j-1}^e = 0, \\ \text{(ii)} & \tilde{W}_j^a (v_j^a - v_{j-1}^e) = 0. \end{array} \quad (8.69)$$

We will show later that these conditions can be satisfied by a suitable choice of the initial vector v_j^a and the initial matrix Z_j^a. If this is the case, Eqs. (8.68) are reduced to

$$- \tilde{Z}_j^a Z_{j-1}^e z_{j-1} + z_j = \tilde{Z}_j^a (v_{j-1}^e - v_j^a). \quad (8.70)$$

Finally, the ansatz (8.61) for the solution of the given BVP on the segment $[\tau_j, \tau_{j+1}]$ is simplified to

$$y_j(x) = Z(x; \tau_j) z_j + v(x; \tau_j). \quad (8.71)$$

At this point, let us assume that after applying the above strategy, we have $w_j = 0$, $j = 0, \ldots, m-1$. Therefore, the solution of the BVP is reduced to the determination of the vectors $z_j \in \mathbb{R}^q$, $j = 0, \ldots, m - 1$. These vectors are determined by Eqs. (8.64) and (8.70) which we can combine to the following system of linear algebraic equations of the dimension mq:

$$\hat{M}^{(m)} z^{(m)} = \hat{q}^{(m)}, \quad (8.72)$$

where

$$\hat{M}^{(m)} \equiv \begin{pmatrix} -\tilde{Z}_1^a Z_0^e & I_q & & & \\ & -\tilde{Z}_2^a Z_1^e & I_q & & \\ & & \ddots & & \ddots \\ & & & -\tilde{Z}_{m-1}^a Z_{m-2}^e & I_q \\ B_a^{(2)} Z_0^a & & & & B_b^{(2)} Z_{m-1}^e \end{pmatrix},$$

$$z^{(m)} \equiv \begin{pmatrix} z_0 \\ \vdots \\ z_{m-2} \\ z_{m-1} \end{pmatrix} \quad \text{and} \quad \hat{q}^{(m)} \equiv \begin{pmatrix} \tilde{Z}_1^a(v_0^e - v_1^a) \\ \vdots \\ \tilde{Z}_{m-1}^a(v_{m-2}^e - v_{m-1}^a) \\ \beta^{(2)} - B_a^{(2)} v_0^a - B_b^{(2)} v_{m-1}^e \end{pmatrix}.$$

Now, from this general concept, special variants of the stabilized march method can be developed by a specific choice of Z_j^a and v_j^a. Two such variants, the *Godunov-Conte method* (see [1, 14]) and an *elimination algorithm* (see [12, 13]), will be described in brief here.

The characteristic feature of the Godunov-Conte method is the continuous orthogonalization of the columns of the matrix of the complementary functions to guarantee their linear independence on each segment.

Similar to the method of complementary functions, in the first step of the Godunov-Conte method, a QR-factorization of the matrix $(B_a^{(1)})^T$ is computed (see formula (8.30)):

$$(B_a^{(1)})^T = Q_0 \begin{pmatrix} U_0 \\ 0 \end{pmatrix} = (Q_0^{(1)} | Q_0^{(2)}) \begin{pmatrix} U_0 \\ 0 \end{pmatrix} = Q_0^{(1)} U_0, \qquad (8.73)$$

where $U_0 \in \mathbb{R}^{p \times p}$, $Q_0^{(1)} \in \mathbb{R}^{n \times p}$, and $Q_0^{(2)} \in \mathbb{R}^{n \times q}$. Thus, $B_a^{(1)} = U_0^T (Q_0^{(1)})^T$. If we set

$$Z_0^a \equiv Q_0^{(2)} \quad \text{and} \quad v_0^a \equiv Q_0^{(1)} U_0^{-T} \beta^{(1)}, \qquad (8.74)$$

then we have

(i) $B_a^{(1)} Z_0^a = U_0^T (Q_0^{(1)})^T Q_0^{(2)} = 0$,   and

(ii) $B_a^{(1)} v_0^a = U_0^T (Q_0^{(1)})^T Q_0^{(1)} U_0^{-T} \beta^{(1)} = \beta^{(1)}$,

i.e., the conditions (8.63) are satisfied by this choice of the initial values.

To fulfill also the conditions (8.69), in the jth step of the stabilized march method (i.e., in the transition from $[\tau_{j-1}, \tau_j]$ to $[\tau_j, \tau_{j+1}]$), a QR-factorization of Z^e_{j-1} is computed:

$$Z^e_{j-1} = Q_j \begin{pmatrix} U_j \\ 0 \end{pmatrix} = (Q_j^{(2)} | Q_j^{(1)}) \begin{pmatrix} U_j \\ 0 \end{pmatrix} = Q_j^{(2)} U_j, \tag{8.75}$$

where $U_j \in \mathbb{R}^{q \times q}$, $Q_j^{(1)} \in \mathbb{R}^{n \times p}$, and $Q_j^{(2)} \in \mathbb{R}^{n \times q}$. Now, setting

$$Z^a_j \equiv Q_j^{(2)}, \quad W^a_j \equiv Q_j^{(1)}, \quad \text{and} \quad v^a_j \equiv Q_j^{(1)} (Q_j^{(1)})^T v^e_{j-1}, \tag{8.76}$$

we have

$$(W^a_j | Z^a_j)^{-1} = (Q_j^{(1)} | Q_j^{(2)})^{-1} = \begin{pmatrix} (Q_j^{(1)})^T \\ (Q_j^{(2)})^T \end{pmatrix} = \begin{pmatrix} \tilde{W}^a_j \\ \tilde{Z}^a_j \end{pmatrix}.$$

The matrix Z^a_j and the vector v^a_j in (8.76) satisfy the conditions (8.69) as the following calculation shows:

(i) $\tilde{W}^a_j Z^e_{j-1} = (Q_j^{(1)})^T Q_j^{(2)} U_j = 0,$

(ii) $\tilde{W}^a_j (v^a_j - v^e_{j-1}) = (Q_j^{(1)})^T \left(Q_j^{(1)} (Q_j^{(1)})^T v^e_{j-1} - v^e_{j-1} \right)$

$\qquad\qquad\qquad = (Q_j^{(1)})^T v^e_{j-1} - (Q_j^{(1)})^T v^e_{j-1} = 0.$

Moreover, we have

$$\tilde{Z}^a_j Z^e_{j-1} = U_j, \quad \tilde{Z}^a_j \left(v^e_{j-1} - v^a_j \right) = (Q_j^{(2)})^T v^e_{j-1},$$

which implies that the system matrix in (8.72) can be written in the simple form

$$\hat{M}^{(m)} \equiv \begin{pmatrix} -U_1 & I_q & & & \\ & -U_2 & I_q & & \\ & & \ddots & \ddots & \\ & & & -U_{m-1} & I_q \\ B_a^{(2)} Z_0^a & & & & B_b^{(2)} Z_{m-1}^e \end{pmatrix}. \tag{8.77}$$

Finally, the right-hand side of this algebraic system takes on the form

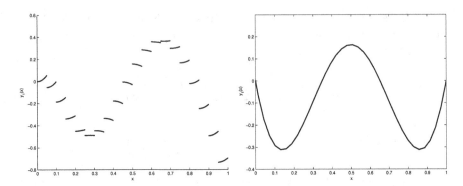

Fig. 8.4 Stabilized march solution of problem (8.38) with $\lambda = 5$ and $m = 20$

$$
\hat{q}^{(m)} \equiv \begin{pmatrix}
(Q_1^{(2)})^T v_0^e \\
\vdots \\
(Q_{m-1}^{(2)})^T v_{m-2}^e \\
\beta^{(2)} - B_a^{(2)} Q_0^{(1)} U_0^{-T} \beta^{(1)} - B_b^{(2)} v_{m-1}^e
\end{pmatrix}. \tag{8.78}
$$

For a direct comparison with the multiple shooting method (see Fig. 8.3), in Fig. 8.4 the initial trajectories and the numerical result of the stabilized march method (Godunov-Conte version) for the test problem (8.38) are presented.

The elimination algorithm of the stabilized march method starts with the following factorization of the matrix $B_a^{(1)}$

$$
B_a^{(1)} = (L_1 \mid 0)\, U\, P = (L_1 \mid 0) \begin{pmatrix} U_1 & H \\ 0 & I_q \end{pmatrix} P. \tag{8.79}
$$

Similarly to (8.33) the initial matrix Z_0^a and the initial vector v_0^a are chosen as

$$
Z_0^a \equiv P^T \begin{pmatrix} -U_1^{-1} H \\ I_q \end{pmatrix} \quad \text{and} \quad v_0^a \equiv P^T \begin{pmatrix} U_1^{-1} L_1^{-1} \beta^{(1)} \\ 0 \end{pmatrix}. \tag{8.80}
$$

It can be easily shown that these quantities satisfy the conditions (8.63). The factorization of matrices Z_{j-1}^e is also not realized by orthogonalization techniques. Rather, at the interior shooting points τ_j, $j = 1, \ldots, m-1$, the following LU-factorization is computed:

$$
Z_{j-1}^e = P_j\, L^{(j)} \begin{pmatrix} U_j \\ 0 \end{pmatrix} \equiv P_j \begin{pmatrix} L_1^{(j)} & 0 \\ H^{(j)} & I_p \end{pmatrix} \begin{pmatrix} U_j \\ 0 \end{pmatrix}, \tag{8.81}
$$

where the permutation matrix $P_j \in \mathbb{R}^{n \times n}$ describes the permutations of the rows, $U_j \in \mathbb{R}^{q \times q}$ is a upper triangular matrix, $L_1^{(j)} \in \mathbb{R}^{q \times q}$ is a lower triangular matrix whose elements are smaller or equal to one in absolute value, $H^{(j)} \in \mathbb{R}^{p \times q}$ is an arbitrary matrix with elements which are bounded in absolute value by one, and $I_p \in \mathbb{R}^{p \times p}$ is the unit matrix. If we set

$$Z_j^a \equiv P_j \begin{pmatrix} L_1^{(j)} \\ H^{(j)} \end{pmatrix}, \tag{8.82}$$

the first equation in (8.69) is satisfied and it holds $\tilde{Z}_j^a Z_{j-1}^e = U_j$. Thus, the matrix of the linear algebraic system (8.72) can be written again in the form (8.77). Finally, if the initial vector v_j^a is determined as solution of the linear system

$$\begin{pmatrix} -H^{(j)}(L_1^{(j)})^{-1} \ I_p \\ I_q \qquad 0 \end{pmatrix} P_j^T v_j^a = \left(\begin{pmatrix} -H^{(j)}(L_1^{(j)})^{-1} \mid I_q \end{pmatrix} P_j^T v_{j-1}^e \\ 0 \right), \tag{8.83}$$

then the second equation in (8.69) is also satisfied.

The reduction of the dimension of the linear algebraic system from mn (multiple shooting method) to mq (both variants of the stabilized march method) is not the decisive advantage of this strategy. More important is the fact that the stabilized march method requires only the solution of $q + 1$ IVPs per segment $[\tau_j, \tau_{j+1}]$, namely,

- 1 IVP for the particular solution $v(x; \tau_j)$:

$$\mathscr{L}v(x; \tau_j) = r(x), \quad v(\tau_j; \tau_i) = v_j^a, \tag{8.84}$$

where

$$v_j^a \text{ is determined by } \begin{cases} (8.74) \text{ or } (8.80), \text{ if } j = 0, \\ (8.76) \text{ or } (8.84), \text{ if } j > 0, \end{cases};$$

- q IVPs for the columns of the matrix $Z(x; \tau_j)$:

$$\mathscr{L}Z(x; \tau_j) = 0, \quad Z(\tau_j; \tau_j) = Z_j^a, \tag{8.85}$$

where

$$Z_j^a \text{ is determined by } \begin{cases} (8.74) \text{ or } (8.80), \text{ if } j = 0, \\ (8.76) \text{ or } (8.82), \text{ if } j > 0 \end{cases}.$$

Thus, we have shown that the question marks in Table 8.5 can be removed, i.e., our guess was correct.

But notice that the stabilized march method is not a parallel algorithm. In the multiple shooting method, the solution of the IVPs on the individual segments can be performed independently since the initial matrices Y_j^a and initial vectors v_j^a are fixed at the beginning of the method. Therefore, the multiple shooting method is well suited for the parallel computing technology and is often called *parallel shooting method*. In contrast, the stabilized march method is a sequential algorithm since the initial values Y_j^a and v_j^a for the integration on $[\tau_j, \tau_{j+1}]$ are constructed on the basis of Y_{j-1}^e and v_{j-1}^e, i.e., they require the result of the integrations on the previous segment $[\tau_{j-1}, \tau_j]$.

Finally, it should be mentioned that the stabilized march method has similar positive stability characteristics as the multiple shooting method (see [13]).

At the Friedrich Schiller University Jena, a program package RWPM has been developed which is based on the theory presented in this chapter (see, e.g., [6–8,17]). This package is also suitable for solving nonlinear BVPs and parametrized nonlinear problems (bifurcation problems). The shooting points τ_j are adapted automatically by checking the growth of the elements in the matrices $Y(x; \tau_j)$ and $Z(x; \tau_j)$. There are implementations of RWPM in FORTRAN and MATLAB.

8.9   Exercises

Exercise 8.1. Given the following BVP

$$y'(x) - \begin{pmatrix} 0 & \lambda \\ \lambda & 0 \end{pmatrix} y(x) = r(x), \quad 0 \le x \le 1,$$

$$y_1(x) = \beta_1, \quad y_1(1) = \beta_2,$$

where $\lambda > 0$ is a real parameter.

1. Determine a fundamental matrix $Y(x; 0)$ of the corresponding homogeneous ODE which satisfies $\mathscr{B}Y(x; 0) = I$.
2. Determine the solution of the given BVP with the method of variation of parameters using the result from 1.
3. Represent the solution of the BVP with Green's function.

Exercise 8.2. Consider the BVP (8.1a) and (8.1b). Let $Y(x; a)$ be a fundamental matrix with

$$\mathscr{L}Y(x; a) = 0, \quad Y(a; a) = I.$$

Show that $Y(x; a)$ has the following properties of an algebraic group:

(1) $Y(x; \tau)Y(\tau; a) = Y(x; a)$.
(2) $(Y(a; \tau))^{-1} = Y(\tau; a)$.

Exercise 8.3. On the interval $[a,b]$, a continuously differentiable function $f : [a,b] \to \mathbb{R}$ has to be approximated by a polynomial $P(x)$ which satisfies for given f_a, f_b, f'_a, and f'_b:

$$P \in \Pi_3[a,b] : \quad P(a) = f_a, \quad P(b) = f_b, \quad P'(a) = f'_a, \quad P'(b) = f'_b,$$

where $f_w = f(w)$, $f'_w = f'(w)$, $w = a,b$, and Π_3 denotes the set of all polynomials whose degree is less or equal 3. This problem has a unique solution.

1. Formulate this interpolation problem as a BVP for a system of first-order ODEs.
2. Develop a MATLAB function hermitbvpssm for the numerical solution of this BVP by the simple shooting method.
3. Develop a MATLAB function hermitbvpmcf for the numerical solution of this BVP by the method of complementary functions.

Hint: Use the existing MATLAB algorithms to solve the associated systems of linear equations. For the solution of the IVPs, apply the MATLAB code ode45.

Exercise 8.4.

1. Let $B \in \mathbb{R}^{p \times n}$, $0 < p \leq n$, be a matrix with rank$(B) = p$. Show how the well-known Gaussian elimination must be modified to realize the following factorization:

$$B = (L \mid 0) \begin{pmatrix} U & H \\ 0 & I_{n-p} \end{pmatrix} P,$$

where $L \in \mathbb{R}^{p \times p}$ is a lower triangular matrix, $U \in \mathbb{R}^{p \times p}$ is a upper triangular matrix, and $P \in \mathbb{R}^{n \times n}$ is a permutation matrix. Moreover, it holds $|u_{ij}| \leq 1$ and $|u_{ii}| = 1$, $i,j = 1,2,\ldots,p$.
2. Implement the above factorization as a MATLAB function

$$[L,U,H,P] = lup(A).$$

Test the function lup with the help of self-generated test matrices A.

Exercise 8.5. Given the linear BVP (8.1a) and (8.1b) with partially separated boundary conditions (8.2).

1. Develop a MATLAB function bvpsep for the solution of this BVP by the method of complementary functions. Use the existing MATLAB code qr to generate the associated QR-factorization. For the solution of the IVPs, apply the MATLAB code ode45.
2. Consider the following spline interpolation problem: On the grid

$$a = x_1 < x_2 < \cdots < x_n < x_{n+1} = b$$

are given the values $f_1, f_2, \ldots, f_{n+1}$ of a function $f(x)$.

We are looking for functions $s_k(x) \in \Pi_3$, $x_k \leq x \leq x_{k+1}$, $k = 1,\ldots,n$, which satisfy

$$s_k(x_k) = f_k, \quad s_k(x_{k+1}) = f_{k+1}, \quad k = 1,\ldots,n,$$

$$s_k^{(j)}(x_{k+1}) = s_{k+1}^{(j)}(x_{k+1}), \quad k = 1,\ldots,n-1, \quad j = 1,2,$$

$$s_1''(x_1) = 0, \quad s_n''(x_{n+1}) = 0.$$

The function $s(x)$ on $[a,b]$ which satisfies $s(x) = s_k(x)$ for $x \in [x_k, x_{k+1})$ is called a *cubic spline function*. The determination of this spline function can be formulated as a BVP for $s(x)$. Formulate this BVP and compute the numerical solution by the code bvpsep. Graphically display $s(x)$, $s'(x)$, and $s''(x)$. In the graphical display of $s(x)$, the fulfillment of the interpolation conditions should be seen.

3. Test the MATLAB function bvpsep for different spline interpolation problems using self-selected functions $f : [a, b] \to \mathbb{R}$.
4. In MAPLE some tools exist to compute the cubic spline functions in closed form. Use these tools to examine the quality of the bvpsep generated splines.
5. Solve the BVP for the spline function with available software tools (MATLAB: bvp4c, rwp; MAPLE: dsolve).

Exercise 8.6. Given the linear BVP (8.1a) and (8.1b).

1. Develop a MATLAB implementation of the standard form of the multiple shooting method where the shooting points are provided by the user.
2. Determine the numerical solution of the BVP (8.38) for various values of the parameter λ. Display the solutions graphically.

Exercise 8.7. Prove formula (8.54).

Exercise 8.8. Consider the BVP (8.1a) and (8.1b) on the interval $[a, b] = [0, 1]$, with

$$A(x) = \begin{pmatrix} 1 & -3 & 3 \\ 3 & -5 & 3 \\ 3 & -3 & 2 \end{pmatrix}, \quad r(x) = \begin{pmatrix} 0 \\ 0 \\ 0 \end{pmatrix},$$

$$B_a = \begin{pmatrix} 1 & 1 & 1 \\ 0 & 1 & 0 \\ 0 & 0 & 1 \end{pmatrix}, \quad B_b = \begin{pmatrix} 0 & 0 & 0 \\ 0 & -1 & 0 \\ 0 & 0 & -1 \end{pmatrix}, \quad \beta = \begin{pmatrix} 1 \\ 1 \\ 1 \end{pmatrix}.$$

1. Show that this BVP has a unique solution.
2. Determine the exact solution.
3. Solve this BVP numerically with the stabilized march method (Godunov-Conte version). Write a MATLAB code where the QR-factorization and the solution of

the linear algebraic system $\hat{M}^{(m)} z^{(m)} = \hat{q}^{(m)}$ are computed by given MATLAB functions. This implementation of the stabilized march method should only be based on two interior shooting points. Represent the numerical solution graphically. Compare the numerical solution with the exact solution.

Exercise 8.9. Prove:

1. The linear BVP

$$u''(x) + u(x) = 0, \quad u(0) = u(\pi) = 1,$$

is unsolvable.

2. The linear BVP

$$u''(x) + u(x) = 0, \quad u(0) = 1, \quad u(\pi) = -1,$$

has an infinite number of solutions.

Exercise 8.10. Has the BVP

$$-y''(x) - 4y'(x) + \sin(x)y(x) = \cos(x), \quad y(0) = y(\pi), \quad y'(0) = y'(\pi),$$

a unique solution?

Exercise 8.11. Compute the solution of the following BVP:

$$y'(x) = \begin{pmatrix} 0 & 1 \\ \lambda^2 & 0 \end{pmatrix} y(x) + (1 - \lambda^2) \begin{pmatrix} 0 \\ e^x \end{pmatrix}, \quad x \in [0, b],$$

$$\begin{pmatrix} 1 & 0 \\ 0 & 0 \end{pmatrix} y(0) + \begin{pmatrix} 0 & 0 \\ 1 & 0 \end{pmatrix} y(b) = \begin{pmatrix} 0 \\ e^b \end{pmatrix}, \quad \lambda > 0, \quad b > 0,$$

in dependence of λ and b.

Exercise 8.12. The linear BVP

$$y''(x) = 100 \, y(x), \quad 0 \le x \le 3,$$

$$y(0) = 1, \quad y(3) = e^{-30},$$

has to be solved by the simple shooting method. For this purpose, compute the solution $u(x; s)$ of the associated IVP

$$u''(x) = 100 \, u(x), \quad 0 \le x \le 3,$$

$$u(0) = 1, \quad u'(0) = s,$$

and try to determine $s = s^*$ such that the equation

$$u(3; s) = e^{-30}$$

is satisfied. How large is the relative error of $u(3; \tilde{s})$ if the numerical approximation $\tilde{s}$ of s^* has the relative error ε?

Exercise 8.13. Use the simple shooting method, the method of complementary functions, the multiple shooting method, and the stabilized march method to solve the following linear boundary value problems. Plot your solutions.

(1) $y''(x) + \dfrac{2}{x} y'(x) + y(x) = 0$, $y(1) = 1$, $y(2) = 5$.

(2) $y''(x) + y'(x) + x\, y(x) = 0$, $y(0) = 1$, $y(1) = 0$.

(3) $y''(x) + \dfrac{2}{x} y'(x) - \dfrac{2}{x^2} y(x) = \dfrac{\sin(x)}{x^2}$, $y(1) = 1$, $y(2) = 2$.

(4) $y''(x) + \sqrt{x}\, y'(x) + y(x) = e^x$, $y(0) = 0$, $y(1) = 0$.

(5) $y''(x) - \dfrac{2x}{1 - x^2}\, y'(x) + \dfrac{12}{1 - x^2}\, y(x) = 0$, $y(0) = 0$, $y(0.5) = 4$.

Exercise 8.14. Determine the exact solution of the BVP:

$$y''(x) + y(x) = x, \quad y(0) = 1, \quad y(\pi/2) = 0.$$

Solve this BVP with the simple shooting method and determine the relative error of the numerical solution.

References

1. Conte, S.D.: The numerical solution of linear boundary value problems. SIAM Rev. **8**, 309–321 (1966)
2. Deuflhard, P., Bornemann, F.: Numerische Mathematik II. Walter de Gruyter, Berlin/New York (2002)
3. Hermann, M.: Zur Stabilität der Mehrzielmethode: Das Keller'sche Verfahren. Tech. Rep. 41, Friedrich Schiller University at Jena (1980)
4. Hermann, M.: Shooting methods for two-point boundary value problems – a survey. In: Hermann, M. (ed.) Numerische Behandlung von Differentialgleichungen, Wissenschaftliche Beiträge der FSU Jena, pp. 23–52. Friedrich-Schiller-Universität, Jena (1983)
5. Hermann, M.: Numerik gewöhnlicher Differentialgleichungen. Anfangs- und Randwertprobleme. Oldenbourg Verlag, München/Wien (2004)
6. Hermann, M., Kaiser, D.: RWPM: a software package of shooting methods for nonlinear two-point boundary value problems. Appl. Numer. Math. **13**, 103–108 (1993)
7. Hermann, M., Kaiser, D.: Shooting methods for two-point BVPs with partially separated endconditions. ZAMM **75**(9), 651–668 (1995)
8. Hermann, M., Kaiser, D.: Numerical methods for parametrized two-point boundary value problems – a survey. In: Alt, W., Hermann, M. (eds.) Berichte des IZWR, vol. Math/Inf/06/03, pp. 23–38. Friedrich-Schiller-Universität Jena, Jenaer Schriften zur Mathematik und Informatik (2003)

9. Lentini, M., Pereyra, V.: An adaptive finite difference solver for nonlinear two- point boundary value problems with mild boundary layers. SIAM J. Numer. Anal. **14**, 91–111 (1977)

10. Morrison, D.D., Riley, J.D., Zancanaro, J.F.: Multiple shooting methods for two point boundary value problems. Commun. ACM **5**(12), 613–614 (1962)

11. Osborne, M.R.: On shooting methods for boundary value problems. J. Math. Anal. Appl. **27**(2), 417–433 (1969)

12. Osborne, M.R.: Aspects of the numerical solution of boundary value problems with separated boundary conditions. Working paper, Mathematical Sciences Institute, Australian National University, Canberra (1978)

13. Osborne, M.R.: The stabilized march is stable. SIAM J. Numer. Anal. **16**, 923–933 (1979)

14. Scott, M.R., Watts, H.A.: Computational solution of linear two-point boundary value problems via orthonormalization. Numer. Anal. **14**, 40–70 (1977)

15. Skeel, R.D.: Iterative refinement implies numerical stability for Gaussian elimination. Math. Comput. **35**(151), 817–832 (1980)

16. Stoer, J., Bulirsch, R.: Introduction to Numerical Analysis. Springer, New York/Berlin/Heidelberg (2002)

17. Wallisch, W., Hermann, M.: Schießverfahren zur Lösung von Rand- und Eigenwertaufgaben. Teubner-Texte zur Mathematik, Bd. 75. Teubner Verlag, Leipzig (1985)

18. Wright, S.J.: Stable parallel elimination for boundary value ODE's. Tech. Rep. MCS-P229-0491, Mathematics and Computer Science Division, Argonne National Laboratory, Argonne(1991)

19. Wright, S.J.: Stable parallel algorithms for two-point boundary value problems. J. Sci. Stat. Comput. **13**, 742–764 (1992)

20. Wright, S.J.: Stable parallel elimination for boundary value odes. Numer. Math. **67**, 521–535 (1994)

Appendix A
Power Series

1. $e^{ax} = 1 + \dfrac{a\,x}{1!} + \dfrac{(a\,x)^2}{2!} + \dfrac{(a\,x)^3}{3!} + \cdots$

2. $a^x = 1 + \dfrac{x\,\ln(a)}{1!} + \dfrac{(x\,\ln(a))^2}{2!} + \dfrac{(x\,\ln(a))^3}{3!} + \cdots, \quad a > 0$

3. $(1 + a\,x)^{-1} = 1 - a\,x + (a\,x)^2 - (a\,x)^3 + \cdots, \quad |a\,x| < 1$

4. $\ln(1 + x) = x - \dfrac{x^2}{2} + \dfrac{x^3}{3} - \dfrac{x^4}{4} + \cdots, \quad |x| < 1$

5. $\sin(x) = x - \dfrac{x^3}{3!} + \dfrac{x^5}{5!} - \dfrac{x^7}{7!} + \cdots$

6. $\cos(x) = 1 - \dfrac{x^2}{2!} + \dfrac{x^4}{4!} - \dfrac{x^6}{6!} + \cdots$

7. $\tan(x) = x + \dfrac{x^3}{3} + \dfrac{2x^5}{15} + \dfrac{17x^7}{315} + \cdots, \quad |x| < \dfrac{\pi}{2}$

8. $\sec(x) = 1 + \dfrac{x^2}{2!} + \dfrac{5x^4}{4!} + \dfrac{61x^6}{6!} + \cdots, \quad |x| < \dfrac{\pi}{2}$

9. $\operatorname{cosec}(x) = \dfrac{1}{x} + \dfrac{x}{6} + \dfrac{7x^3}{360} + \dfrac{13x^5}{5{,}040} + \cdots, \quad |x| < \pi$

10. $\tan^{-1}(x) = x - \dfrac{x^3}{3} + \dfrac{x^5}{5} - \dfrac{x^7}{7} + \cdots$

11. $\sinh(x) = x + \dfrac{x^3}{3!} + \dfrac{x^5}{5!} + \dfrac{x^7}{7!} + \cdots$

12. $\cosh(x) = 1 + \dfrac{x^2}{2!} + \dfrac{x^4}{4!} + \dfrac{x^6}{6!} + \cdots$

13. $\tanh^{-1}(x) = x + \dfrac{x^3}{3} + \dfrac{x^5}{5} + \dfrac{x^7}{7} + \cdots$

M. Hermann and M. Saravi, *A First Course in Ordinary Differential Equations: Analytical and Numerical Methods*, DOI 10.1007/978-81-322-1835-7, © Springer India 2014

Appendix B
Some Elementary Integration Formulas

1. $\int x^n dx = \dfrac{x^{n+1}}{n+1} + c, \quad n \neq -1$

2. $\int u'(x)\, u(x)^n dx = \dfrac{u(x)^{n+1}}{n+1} + c, \quad n \neq -1$

3. $\int \dfrac{1}{x} dx = \ln(x) + c$

4. $\int e^{ax} dx = \dfrac{1}{a} e^{ax} + c$

5. $\int a^x dx = \dfrac{1}{\ln(a)} a^x + c$

6. $\int x^n \ln(x) dx = x^{n+1} \left[\dfrac{1}{n+1} \ln(x) - \dfrac{1}{(n+1)^2} \right] + c, \quad n \neq -1$

7. $\int x \ln(x) dx = x^2 \left(\dfrac{1}{2} \ln(x) - \dfrac{1}{4} \right) + c$

8. $\int \sin(a x) dx = -\dfrac{1}{a} \cos(a x) + c$

9. $\int \cos(a x) dx = \dfrac{1}{a} \sin(a x) + c$

10. $\int \tan(a x) dx = -\dfrac{1}{a} \ln(\cos(a x)) + c$

11. $\int \cot(a x) dx = \dfrac{1}{a} \ln(\sin(a x)) + c$

12. $\int \sec(a x) dx = \dfrac{1}{a} \ln\big(\sec(a x) + \tan(a x) \big) + c$

13. $\int \operatorname{cosec}(a x) dx = \dfrac{1}{a} \ln\big(\operatorname{cosec}(a x) - \cot(a x) \big) + c$

M. Hermann and M. Saravi, *A First Course in Ordinary Differential Equations: Analytical and Numerical Methods*, DOI 10.1007/978-81-322-1835-7, © Springer India 2014

14. $\displaystyle\int \sinh(a\,x)dx = \frac{1}{a}\cosh(a\,x) + c$

15. $\displaystyle\int \cosh(a\,x)dx = \frac{1}{a}\sinh(a\,x) + c$

16. $\displaystyle\int \tanh(a\,x)dx = \frac{1}{a}\ln(\cosh(a\,x)) + c$

17. $\displaystyle\int \coth(a\,x)dx = \frac{1}{a}\ln(\sinh(a\,x)) + c$

18. $\displaystyle\int \text{sech}(a\,x)dx = \frac{1}{a}\ln(\sinh(a\,x)) + c$

19. $\displaystyle\int \text{cosech}(a\,x)dx = \frac{1}{a}\ln\left(\tanh\left(\frac{x}{2}\right)\right) + c$

20. $\displaystyle\int \frac{dx}{\sqrt{a^2 - x^2}} = \sin^{-1}\left(\frac{x}{a}\right) + c$

21. $\displaystyle\int \frac{dx}{\sqrt{x^2 - a^2}} = \sinh^{-1}\left(\frac{x}{a}\right) + c$

22. $\displaystyle\int \frac{dx}{\sqrt{x^2 + a^2}} = \ln\left(x + \sqrt{x^2 + a^2}\right) + c$

23. $\displaystyle\int \frac{dx}{a^2 + x^2} = \frac{1}{a}\tan^{-1}\left(\frac{x}{a}\right) + c$

24. $\displaystyle\int \frac{dx}{a^2 - x^2} = \frac{1}{2a}\ln\left(\frac{x + a}{x - a}\right) + c$

25. $\displaystyle\int \frac{dx}{x\sqrt{x^2 - a^2}} = \frac{1}{a}\sec^{-1}\left(\frac{x}{a}\right) + c$

26. $\displaystyle\int \frac{dx}{x\sqrt{x^2 + a^2}} = \frac{1}{a}\ln\left(\frac{x}{a + \sqrt{x^2 + a^2}}\right) + c$

27. $\displaystyle\int \frac{dx}{x\sqrt{x^2 - a^2}} = \frac{1}{a}\ln\left(\frac{x}{a + \sqrt{x^2 - a^2}}\right) + c$

28. $\displaystyle\int e^{a\,x}\sin(b\,x)dx = \frac{e^{a\,x}}{a^2 + b^2}\left(a\sin(b\,x) - b\cos(b\,x)\right) + c$

29. $\displaystyle\int e^{a\,x}\cos(b\,x)dx = \frac{e^{a\,x}}{a^2 + b^2}\left(a\cos(b\,x) + b\sin(b\,x)\right) + c$

30. $\displaystyle\int \frac{\sin(x)}{x}dx = \sum_{n=1}^{\infty} \frac{(-1)^{n-1}x^{2n-1}}{(2n - 1)(2n - 1)!} + c$

31. $\displaystyle\int \frac{1 - \cos(x)}{x}dx = \sum_{n=1}^{\infty} \frac{(-1)^{n-1}x^{2n}}{(2n)(2n)!} + c$

Appendix C
Table of Laplace Transforms

$f(x)$	$F(s)$
k	$\dfrac{k}{s}, \quad s > 0$
e^{ax}	$\dfrac{1}{s-a}, \quad s > a$
$\sin(a\,x)$	$\dfrac{a}{s^2 + a^2}$
$\cos(a\,x)$	$\dfrac{s}{s^2 + a^2}$
$x^n, \quad n = 1, 2, \ldots$	$\dfrac{n!}{s^{n+1}}$
$x^n\,e^{ax}, \quad n = 1, 2, \ldots$	$\dfrac{n!}{(s-a)^{n+1}}$
$e^{ax}\sin(b\,x)$	$\dfrac{b}{(s-a)^2 + b^2}$
$x^{-\frac{1}{2}}$	$\sqrt{\pi/s}$
$f(a\,x)$	$\dfrac{1}{a}\,F\left(\dfrac{s}{a}\right)$
$e^{ax}\,f(x)$	$F(s-a)$
$\mathrm{erf}(\sqrt{a\,x})$	$\dfrac{\sqrt{a}}{s\sqrt{s+a}}$
$e^{ax}\,\mathrm{erf}(\sqrt{a\,x})$	$\dfrac{\sqrt{a}}{(s-a)\sqrt{s}}$

(continued)

M. Hermann and M. Saravi, *A First Course in Ordinary Differential Equations: Analytical and Numerical Methods*, DOI 10.1007/978-81-322-1835-7, © Springer India 2014

(continued)

$f(x)$	$F(s)$
$\dfrac{\cos(2\sqrt{a\,x})}{\sqrt{\pi\,x}}$	$\dfrac{e^{-\frac{a}{s}}}{\sqrt{s}}$
$\dfrac{\sin(2\sqrt{a\,x})}{\sqrt{\pi\,x}}$	$\dfrac{e^{-\frac{a}{s}}}{s\sqrt{s}}$
$\dfrac{e^{-\frac{a^2}{4x}}}{\sqrt{\pi\,x}}$	$\dfrac{e^{-a\,s}}{\sqrt{s}}$
$\ln(x)$	$\dfrac{-(\gamma + \ln(s))}{s}, \quad \gamma = 0.5772156\ldots$
$x^n \ln(x)$	$\dfrac{\Gamma'(n+1) - \Gamma(n+1)\ln(s)}{s^{n+1}}, \quad n > -1$

Index

Printed in the United States
By Bookmasters